# Resolution Proof Systems

# Automated Reasoning Series

VOLUME 4



***The titles published in this series are listed at the end of this volume.***

# Resolution Proof Systems

## An Algebraic Theory

by

ZBIGNIEW STACHNIAK

*York University,*
*North York, Canada*

Kluwer Academic Publishers

Dordrecht / Boston / London

A C.I.P. Catalogue record for this book is available from the Library of Congress.

ISBN-13:978-94-010-7251-9 e-ISBN-13:978-94-009-1677-7
DOI:10.1007/978-94-009-1677-7

Published by Kluwer Academic Publishers,
P.O. Box 17, 3300 AA Dordrecht, The Netherlands.

Kluwer Academic Publishers incorporates
the publishing programmes of
D. Reidel, Martinus Nijhoff, Dr W. Junk and MTP Press.

Sold and distributed in the U.S.A. and Canada
by Kluwer Academic Publishers,
101 Philip Drive, Norwell, MA 02061, U.S.A.

In all other countries, sold and distributed
by Kluwer Academic Publishers Group,
P.O. Box 322, 3300 AH Dordrecht, The Netherlands.

*Printed on acid-free paper*

*to the memory of my teachers and friends*

Bogusław Iwanuś
and
Tadeusz Kubiński

# CONTENTS

# PREFACE

Automated Reasoning is a research field concerned with the development and application of computer programs that can perform a variety of reasoning tasks ranging from the verification of logical properties (e.g. satisfiability, consistency, theoremhood) to the construction of logical and non-logical objects with desired properties (e.g. proofs, models, counter-models, plans, explanations, programs). Programs that 'reason' can use a variety of proof construction methods; they can be constructed in an *ad hoc* manner or within a general theoretical framework specifically created for such purposes. They can model 'classical' or 'non-classical' reasoning.

This book deals with automated reasoning programs in the context of non-classical reasoning. It presents an algebraic theory of resolution proof systems and shows how this formal algebraic framework can be used for the construction and analysis of resolution based automated reasoning programs.

In the summer of 1963, J. A. Robinson formulated his celebrated resolution rule, a single inference rule on which a refutational counterpart of classical first-order logic can be based. Since that time, the resolution rule, or, as it is often called, the resolution principle, has attracted considerable attention of the Automated Reasoning community. The resolution rule has been subjected to scrupulous analysis, numerous refinements, and sophisticated generalizations in the context of not only classical logic but also many non-classical logical calculi.

If there is one key idea on which this book hinges upon, it is the observation that the resolution principle can be interpreted and formulated in such a manner that resolution proof systems, i.e., refutational deductive proof systems based on the resolution rule, become finite algebraic structures called resolution algebras. A finite (universal) algebra consists of a nonvoid finite set $A$ of elements (the *base set* or *universe*), and of a set of operations acting on $A$. When the elements of such an algebra are well-formed formulas of an object language, and when the algebra is supplemented by a list of unary relations on $A$, a *resolution algebra* is created. It turns out that resolution algebras hold all the information required for the construction of resolution proof systems. To rephrase this metaphorically: resolution proof systems can be conveniently packaged into resolution algebras, and, hence, can be studied in depth and with the precision of tools available to modern universal algebra.

Although the theory of resolution proof systems presented in this book is developed within the algebraic framework, the motivations underlying the central notion of this book, the notion of a resolution principle, are not algebraic and are not driven by the desire to 'algebraize' the area of Automated Reasoning (although algebraic methods have already been found beneficial in Automated Reasoning and other areas of computer science). The proposed interpretation of the principle is rooted in the logical tradition represented by names such as Leśniewski, Łukasiewicz, Tarski, as well as Couturat, Post, and Schröder. It is based on a particular interpretation of the resolution principle as the *rule of consistency verification*: given a logical calculus $\mathcal{P}$, we search for a finite set $Ver$ of formulas of $\mathcal{P}$ such that

(ver) *the consistency status of finite sets of formulas can be verified using $Ver$, i.e., if $X$ is an arbitrary finite and consistent set of formulas of $\mathcal{P}$ and if $v_0, \ldots, v_n$ is an enumeration of the formulas of $Ver$, then for every choice $\alpha_0(p), \ldots, \alpha_n(p)$ of (not necessarily different) formulas of $X$, one of the sets $X \cup \{\alpha_0(p/v_0)\}, \ldots, X \cup \{\alpha_n(p/v_n)\}$ is also consistent.*

The formulas $v_0, \ldots, v_n$ are called *verifiers* as their role is to verify, or to 'witness', the consistency status of sets of formulas.

For some logics, the notion of a verifier coincides with the notion of a formula defining a truth-value. For classical propositional logic, any two formulas $T$ and $F$, that define the truth-values *truth* and *falsehood* respectively, will suffice, and the rule of consistency verification can be rewritten as the inference rule

$$\frac{\alpha_0(p), \alpha_1(p)}{\alpha_0(p/F) \vee \alpha_1(p/T)}.$$

This form of the resolution principle, known as the non-clausal resolution rule, was studied, among others, by Manna and Waldinger (cf. [30]) and Murray (cf. [39]). In general, however, the process of selecting verifiers goes beyond the search for formulas defining truth-values. Verifiers do not always define truth-values; they only define semantic entities that can be used for consistency verification. Indeed, the rule of consistency verification for the 3-valued Łukasiewicz logic cannot be based on $T$ and $F$ alone. The formulas $F$ and $T$ must be supplemented by other verifiers; none of these formulas, however, can define the third truth-value of this logic, intermediate between truth and falsehood.

The term 'verifier' was proposed in 1920s by Tadeusz Kotarbiński in connection with the *rule of verification* formulated by Jan Łukasiewicz in 1921. This rule allows the addition of a new thesis $\alpha$ to the system of propositional calculus, provided that the system already includes theses resulting from $\alpha$ by substituting the respective formulas $F$ and $T$ for one of its propositional variables (cf. [26]). In 1922, Stanisław Leśniewski adopted the rule of verification in his system of Protothetic (later replaced with the extensionality directive). Protothetic is a propositional calculus; however, unlike classical propositional logic, it contains definitional rules for the introduction of variables and constant functors of arbitrary propositional types ('semantic categories'), and allows variables to be bound by universal quantifiers (cf. [59]). The rule of verification permitted Leśniewski to add to Protothetic a new thesis $\forall p[\alpha(p)]$, beginning with a universal quantifier $\forall p$ governing the variable $p$ of any semantic category, when the system already included all theses resulting from $\alpha(p)$ by substituting for $p$ *verifiers* – certain formulas defined for the category of $p$ in advance. Łukasiewicz's rule, restricted to the category of propositions (the basic category of Protothetical expressions), can be expressed by the formula

(a) $$\alpha(p/F) \wedge \alpha(p/T) \rightarrow \forall p[\alpha(p)],$$

where $\alpha(p)$ and $p$ are propositions, $p$ being a variable. In (a), $F$, defined as $\forall p[p]$, and $T$, defined as $\forall p[p \equiv p]$, are the verifiers of the category of propositions. If we let $\exists p[\beta]$ stand for $\neg\forall p[\neg\beta]$, then (a) can be rewritten as

(b) $$\exists p[\alpha(p)] \rightarrow \alpha(p/F) \vee \alpha(p/T).$$

Furthermore, if we let $\alpha$ be $\alpha_0(p) \wedge \alpha_1(p)$, then, on the ground of Protothetic, (b) infers

(c) $$\exists p[\alpha_0(p) \wedge \alpha_1(p)] \rightarrow (\alpha_0(p/F) \vee \alpha_1(p/T)),$$

which, rewritten as an inference rule, expresses the classical non-clausal resolution principle

$$\frac{\alpha_0(p), \alpha_1(p)}{\alpha_0(p/F) \vee \alpha_1(p/T)}.$$

The notion of a verifier that appears in (ver) can be considered a refutational counterpart of the Łukasiewicz-Leśniewski notion of a verifier formulated for non-classical logics.

The laws (a) and (b) were discussed very early in the development of modern logic as they clearly reflect the bivalent nature of classical logic (cf. [9, 57]). In his doctoral dissertation submitted to the University of Warsaw in 1922, Alfred Tarski proved that if $f$ is a truth-function, then the formulas:

(d) $$\forall p[f(p)] \equiv f(p/F) \wedge f(p/T),$$

(e) $$\exists p[f(p)] \equiv f(p/F) \vee f(p/T),$$

as well as their generalizations to extensional propositional functions of more than one argument, are theorems of his quantified propositional calculus (cf. [72]). The laws (d) and (e) constitute the so-called *first law on the bounds of a function* (cf. [9, 57, 72]). In [72] Tarski remarks that

> Łukasiewicz called attention to the fact that the attribution of properties (d) and (e) to every function, or analogous properties to functions of several arguments ... allows considerable simplifications during the construction of the system of logistic.

The study of the resolution principle initiated by J. A. Robinson showed how far reaching and significant these simplifications can be in the context of refutational automated theorem proving methods.

It was Adolf Lindenbaum who observed that the logical connectives of a propositional language $\mathcal{L}$ act as operations on formulas of this language: if $f$ is an $n$-ary connective and $\alpha_1, \ldots, \alpha_n$ are formulas of $\mathcal{L}$, then the formula $f(\alpha_1, \ldots, \alpha_n)$ is the value of $f$ applied to the arguments $\alpha_1, \ldots, \alpha_n$. Hence, $\mathcal{L}$ is a universal algebra, known as *the algebra of formulas.* It was also Lindenbaum who proved, informally speaking, that for every propositional logic $\mathcal{P}$ there exists a denumerable *algebra of truth-values* that provides an adequate semantics for $\mathcal{P}$. And what about verifiers? Is there an algebraic interpretation of the resolution principle? In the following chapters of this book we introduce and discuss the notion of an *algebra of verifiers* and show how these algebras can be used in the theoretical studies and for the effective construction of resolution proof systems. This brings us back to the discussion on the algebraic connections between logic and Automated Reasoning. The main purpose of this book is to explore these connections and to use the findings to guide the construction of efficient automated reasoning programs.

To make this preface into a more accurate account of the book's contents, we must touch upon at least some issues in the area of automated reasoning for non-classical logics that this book is not about. It is not the intention of this book to provide a handbook-like account on the state of affairs in non-classical automated reasoning (for some classes of logical calculi this is taken care of in [19, 75]). Nor it is to provide a translation of other approaches to non-classical automated deduction into the algebraic framework presented in this book. This does not mean, however, that other approaches are less important or less suitable for efficient implementation. On the contrary, in the context of non-classical reasoning, the resolution approach is neither dominant, historically most significant, nor it is destined to universally support all forms of non-classical reasoning. Among the most important rivals to resolution there are varieties of matrix, natural deduction, tableaux, and sequent-based automated proof methods (cf. [3, 12, 13, 16, 19, 41, 46, 48, 75]), there are interpretation and syntax transformation methods (cf. [14, 37, 44, 48, 74]), there are algebraic approaches (cf. [4, 8, 64]), hybrid approaches (cf. [14, 22]), etc.

Even the terms 'resolution rule' and 'resolution principle' do not have unique meanings in the automated reasoning literature, whether or not they are formulated for a specific logical system or a specific class of logical calculi. For instance, among the different proposals for the multiple-valued form of the resolution principle, there are 'clausal' and 'non-clausal' variants, there are strictly syntactic formulations of the principle and formulations that rely on some semantic information being incorporated into its description. There are forms of the principle closely tailored to specific properties of logical calculi (such as the existence of specific logical connectives) and there are general formulations of the principle valid for any logic in a specific class of multiple-valued logics, e.g. for all finitely-valued logics (cf. [2, 19, 24, 38, 40, 45, 64]).

It is also not the intention of this book to evaluate, compare, or classify the many forms of the resolution principle. Automated Reasoning has not yet developed methodological tools for the comparison of different automated reasoning paradigms. This is particularly true in the domain of non-classical logics where automated reasoning methods are frequently inseparable from additional speed-up techniques whose task is to increase the efficiency of the deductive process. It is therefore frequently not clear what constitutes the core of a particular automated reasoning method, and, hence, it is far from obvious what 'yard stick' should be used to measure the performance of particular reasoning programs. In conclusion, this book is a monographic presentation of a certain theory of resolution proof systems. The important but difficult task of the comparison of diverse automated reasoning methods in the domain of non-classical logics lies outside the scope of this book.

## ACKNOWLEDGMENTS

My view of automated reasoning has been greatly enriched by discussions with many people during the period of this book's development. In this regard, I would particularly like to thank Patrick Doherty, Rainer Hähnle, Eric Harley, Neil Murray, Peter O'Hearn, and Erik Rosenthal. I have received useful comments and suggestions on the early drafts of the book from Alexander Leitsch, Grzegorz Malinowski, Giovanni Panti, and David Tremaine. For support of the research behind this book, I thank the Natural Science and Engineering Council of Canada. Special thanks go to Polly Margules, the Acquisition Editor at Kluwer Academic Publishers, for providing valuable support and guidance, and to my wife Ewa for her support and suggestions for numerous improvements in the presentation of the text.

And now, we begin.

# 1

# LOGICAL PRELIMINARIES

In this chapter we briefly discuss the notion of a logical system and look at the refutational principle – an important prerequisite to refutational theorem proving methods. Then we review some of the facts concerning the syntax and the semantics of propositional and first-order logics.

In this book we study logical systems and their resolution proof systems using the tools of abstract (or universal) algebra. We have found the algebraic apparatus to be expressive, rigorous, and sufficiently general to define the required logical and algorithmic notions. The books of Malinowski [29], Rasiowa [52], and Wójcicki [79] exploit the algebraic framework for the study of logical calculi, and so, to a certain extent, do the books of Hähnle [19] and Thistlewaite et al. [73] which deal with automated reasoning techniques in the domain of multiple-valued and relevant logics. In this book we shall use only the basic notions of universal algebra and apply only the elementary results from this area. We shall not review any of them, however, for reasons analogous to those why we shall not review any basic notions from, say, 'naive' set theory – such a task would require a large amount of extra space, the notions are standard and are discussed in depth in the literature. In short, we assume some familiarity with universal algebra, set theory, symbolic logic, and with the computational complexity theory.

For more a complete treatment of the material of this chapter, the reader can consult: the books by Wójcicki [79] and Malinowski [29] for everything pertaining to matrix semantics for propositional logics, and the books by Grätzer [17] and Hpocroft and Ullman [21] for useful background material on universal algebra, formal languages, and computational complexity.

## 1.1 LOGICAL SYSTEMS

Automated Reasoning is a research field concerned with the development and application of computer programs which can perform a variety of reasoning tasks in areas where logic and logic based methodologies are used as formal instruments of scientific investigation. These reasoning tasks are expressed using formal logical frameworks (logical calculi) carefully tailored to represent and to model concepts, relationships, situations, and other objects that may constitute the domain of discourse. Although there are numerous logical systems and many different formalisms that describe them, there are two elements which typically occur in any well-constructed logical system. These are:

(i) a class of meaningful expressions and a grammar for the construction (or an algorithm for the identification) of such expressions;

(ii) a class of logical objects (rules, operations, model structures, etc.) which are used to define the notion of logical entailment.

While (i) describes, roughly speaking, the syntax of the language of a logical system $\mathcal{P}$, (ii) provides $\mathcal{P}$ with an inference engine. For the purpose of our present discussion, we can identify meaningful expressions with well-formed formulas and logical objects mentioned in (ii) with the definition of a logical consequence. Hence, (i) and (ii) give rise to the following definition. A *logic* is a pair

$$\mathcal{P} = \langle L, C \rangle.$$

The first element, $L$, of $\mathcal{P}$ is an infinite and recursive set of *well-formed formulas* (or *formulas*, for short). We assume that formulas are finite strings over a finite alphabet. The way a grammar for $L$ is set (i.e., the way formulas of $L$ are constructed) depends on a particular logic and is of limited interest to us at this place. Since $L$ is recursive, there is an effective way of deciding whether or not a finite string of alphabet symbols constitutes a well-formed formula. We call the number of symbols constituting a formula $\alpha$ the *length* of $\alpha$ and denote it by $length(\alpha)$.

The second element, $C$, of $\mathcal{P}$ is called the *consequence operation* of $\mathcal{P}$. $C$ is a function from the power set $2^L$ of $L$ into $2^L$ and specifies the class of inferences valid in $\mathcal{P}$. We assume that $C$ satisfies the following three conditions. For every $X, Y \subseteq L$:

(c1) $X \subseteq C(X)$; *(inclusion)*

(c2) $X \subseteq Y$ implies $C(X) \subseteq C(Y)$; *(monotonicity)*

(c3) $C(X) = C(C(X))$. *(idempotence)*

For every $X \subseteq L$, $C(X)$ constitutes the set of all consequences of $X$; $\alpha \in C(X)$ can be read as '$\alpha$ is a consequence of $X$ in $\mathcal{P}$' or as '$\alpha$ can be inferred from $X$ in $\mathcal{P}$'. Many important logical notions can be defined in terms of $L$ and $C$. For instance, we can define a *theorem* (or a *tautology*) to be any formula that belongs to $C(\emptyset)$, while the inferential equivalence of two formulas $\alpha$ and $\beta$ can be expressed by the equality $C(\{\alpha\}) = C(\{\beta\})$. Furthermore, by adopting to our framework the method described by Łoś and Suszko in [25], we can define a Hilbert-style axiom system for $\mathcal{P}$ (i.e., a set of axioms and rules of inference which defines $C$), using only $L$ and $C$ as primitives.

We illustrate the definition of a logic using the following example.

EXAMPLE 1.1: Let $A = \{a_0, \ldots, a_n\}$ be an alphabet and let $L$ be any infinite recursive language over $A$. Let $C : 2^L \longrightarrow 2^L$ be the function defined as follows. For every $X \subseteq L$ and every $\alpha \in L$,

(a) $\alpha \in C(X)$ *iff* $\alpha$ is a substring of some $\beta \in X$.

For instance, if $A = \{0, 1\}$ and if $L = \{0^n 1^n : n > 0\}$ (here, $0^n 1^n$ denotes the string consisting of $n$ 0's followed by $n$ 1's), then for every $X \subseteq L$,

$$C(X) = \begin{cases} L, & \text{if } X \text{ is infinite,} \\ \{0^n 1^n : n \leq k\}, & \text{otherwise,} \end{cases}$$

where $k$ is the largest integer such that $0^k 1^k \in X$.

The operation $C$, as defined in (a), is a consequence operation on $L$. Indeed, (c1) and (c2) follow directly from the definition of $C$. To show idempotence, let $X$ be an arbitrary subset of $L$. By inclusion and monotonicity, $C(X) \subseteq C(C(X))$. To show that also $C(C(X)) \subseteq C(X)$, let $\alpha \in C(C(X))$. There are $\beta \in C(X)$ and $\gamma \in X$ such that $\alpha$ is a substring of $\beta$, and $\beta$ is a substring of $\gamma$. So, $\alpha$ is a substring of $\gamma$ and $\alpha \in C(X)$, as required. ∎

In conclusion, in this book we view a logic not as a set of formulas (theses) closed under a fixed set of inference rules but rather as a set of inferences of the form $\langle \alpha, X \rangle$, where $\alpha \in C(X)$. This approach, which represents the so-called inferential direction in Symbolic Logic, is also reflected in the study of non-monotonic inference systems in Artificial Intelligence; the notion of an inference operation proved to be a convenient tool for exploring reasoning processes which are based on the principles of reasoning weaker than the law (c2) of monotonicity (see [15, 23, 27, 28, 68]; see also Chapter 8).

## 1.2 REFUTATIONAL PRINCIPLE

Having the notion of a logic at our disposal, we can look at some of the goals of Automated Reasoning more closely. If $\langle L, C\rangle$ is a logic and $X \cup \{\alpha\}$ is a set of formulas, then one would like to know whether or not $\alpha \in C(X)$. Ideally, what we would like to have is an algorithm $\mathcal{A}$ which, for every formula $\alpha \in L$ and every finite set $X \subseteq L$, outputs 'true' or 'false' in accordance with the following definition.

$$(\Delta) \qquad \mathcal{A}(X,\alpha) = \begin{cases} true, & \text{if } \alpha \in C(X), \\ false, & \text{if } \alpha \notin C(X). \end{cases}$$

In other words, the algorithm $\mathcal{A}$ returns 'true' on input $\langle X, \alpha\rangle$ if and only if $\alpha$ is a consequence of $X$. We call an algorithm $\mathcal{A}$ satisfying $(\Delta)$ an *automated reasoning algorithm* for $C$. Clearly, the existence of automated reasoning algorithms as well as their computational properties depend on a given logic. For some logics such algorithms cannot be found at all. For those logical calculi for which automated reasoning algorithms do exist, the challenge is to find algorithms with the best properties from the standpoint of computational complexity and intended applications.

The notion of consistency is among the most important concepts in Symbolic Logic. Following Post and Tarski, we define a set of formulas $X$ to be *consistent* in a logic $\langle L, C\rangle$ (or $C$-*consistent*), if the set of its consequences does not contain all the formulas, i.e., if $C(X) \neq L$. Let us note that this definition of consistency does not rely on the existence of any special logical symbols in the language (such as negation or a constant representing falsehood). A set $X \subseteq L$ is called $C$-*inconsistent* if it is not $C$-consistent.

The existence of an algorithm which determines the consistency status of finite sets of formulas, i.e., of an algorithm $\mathcal{A}_{con}$ such that for every finite $X \subseteq L$,

$$(\Delta_0) \qquad \mathcal{A}_{con}(X) = true \ \textit{ iff } \ C(X) = L,$$

is one of the prerequisites to refutational theorem proving methods. Again, the existence of algorithms satisfying $(\Delta_0)$ depends on a particular logic.

For many logical calculi, the existence of automated reasoning algorithms and the existence of algorithms for consistency checking are interrelated. For example, in classical propositional logic we can verify whether or not a formula $\alpha$ is a consequence of a finite set $X$ of formulas by determining the consistency status of $X \cup \{\neg\alpha\}$. Indeed, in this logic, $\alpha$ is a consequence of $X$ if and only if $X \cup \{\neg\alpha\}$ is inconsistent. Therefore, if $\mathcal{A}_{con}$ is an algorithm satisfying ($\Delta_0$), then $\mathcal{A}(X, \alpha) \stackrel{df}{=} \mathcal{A}_{con}(X \cup \{\neg\alpha\})$ satisfies ($\Delta$). This construction is the essence of 'proofs by contradiction' in classical logic.

The idea of expressing logical entailment in terms of inconsistency, as it has just been explained using the example of classical logic, lies at the core of refutational theorem proving methods (such as resolution). If $\mathcal{A}_{con}$ is a consistency checking algorithm for a logic $\mathcal{P} = \langle L, C \rangle$, then to determine the validity of inferences in $\mathcal{P}$ we may try the following steps. First, we look for an algorithm that converts (translates) every finite set $X$ of formulas and every formula $\alpha$ into a finite set $X_\alpha$ of formulas so that

(ref) $$\alpha \in C(X) \quad \textit{iff} \quad C(X_\alpha) = L,$$

that is, $\alpha$ is a consequence of $X$ if and only if $X_\alpha$ is inconsistent. If such an algorithm can be constructed, then by ($\Delta_0$) and (ref), $\mathcal{A}(X, \alpha) \stackrel{df}{=} \mathcal{A}_{con}(X_\alpha)$ is an automated reasoning algorithm for $C$. We call $X_\alpha$ the *deduced set* and (ref) the *refutational principle*. It is the refutational principle that provides the link between the direct and indirect (i.e., refutational) theorem proving methods.

The existence of $\mathcal{A}_{con}$ and the satisfaction of (ref) does not always offer an obstacle-free ride towards an efficient refutational theorem proving method. The computational complexity of the automated reasoning algorithm $\mathcal{A}(X, \alpha) \stackrel{df}{=} \mathcal{A}_{con}(X_\alpha)$ depends on:

(I) computational complexity of the construction of $X_\alpha$,

(II) computational complexity of $\mathcal{A}_{con}$.

Both (I) and (II) can be major obstacles towards an efficient implementation. Let us consider (I) first, limiting our discussion to time complexity only. Let $T(n)$ be a total recursive function. We say that $\mathcal{P}$ is $T(n)$-*decidable* if there is a $T(n)$-time bounded automated reasoning algorithm $\mathcal{A}$ for $C$. (In this definition we identify $\mathcal{A}$ with a suitable Turing machine that accepts $\{\langle X, \alpha \rangle : \alpha \in C(X)$ and $X$ is finite$\}$ and $n$ is the sum of lengths of all the formulas of $X \cup \{\alpha\}$, cf. [21].) Informally speaking, $\mathcal{P}$ is $T(n)$-*decidable* if for every finite set $X \cup \{\alpha\} \subseteq L$, $\mathcal{A}$ requires $\leq T(n)$ steps to determine whether or not $\alpha$ is a consequence of $X$ in $\mathcal{P}$.

THEOREM 1.2: *Let $\mathcal{P}$ be a $T(n)$-decidable logic with a finite inconsistent set $X_f$. Let $k$ be the sum of lengths of all the formulas of $X_f$. Then, there exists a $(T(n)+k)$-time bounded algorithm for computing deduced sets $X_\alpha$.*

*Proof:* The proof of this theorem is left as Exercise 1.1. ∎

For many logics the sets $X_\alpha$ satisfying (ref) can be constructed in polynomial time. For example, for classical propositional logic these sets can always be constructed in $n+1$ steps, as we can simply let $X_\alpha$ be $X \cup \{\neg\alpha\}$. Although, by Theorem 1.2, every decidable logic satisfies the refutational principle, there are logical calculi for which time complexity of transforming $X \cup \{\alpha\}$ into $X_\alpha$ is non-polynomial and can be of the same order as time complexity of the best decision procedures for these logics.

THEOREM 1.3: *There exists a decidable logic $\mathcal{P}$ such that every algorithm for constructing deduced sets $X_\alpha$ is of non-polynomial time complexity.*

*Proof:* The proof of this theorem is left as Exercise 1.2. ∎

Now, let us turn to (II). If a logic $\mathcal{P} = \langle L, C\rangle$ satisfies the refutational principle, then the validity of an inference $\alpha \in C(X)$ ($X$ is finite) can be determined by testing the consistency status of $X_\alpha$. The successful construction of an efficient consistency checking algorithm for $\mathcal{P}$ (assuming that such an algorithm exists) depends on whether the definition of $\mathcal{P}$ which is available to us can provide enough insights into the algorithmic properties of $C$. In some cases alternative formalizations or refutational variants of $\mathcal{P}$ have to be sought. These refutational variants are logical calculi which have the same inconsistent sets of formulas as $\mathcal{P}$. Let us suppose that $\mathcal{P}_1 = \langle L, C_1\rangle$ is a refutational variant of $\mathcal{P}$, i.e., let us assume that for every $X \subseteq L$,

$$(\Delta_1) \qquad C(X) = L \textit{ iff } C_1(X) = L.$$

By $(\Delta_1)$, every consistency checking algorithm for $\mathcal{P}_1$ will work for $\mathcal{P}$ as well. Indeed, if $\mathcal{A}_{con}$ is one of such algorithms and if $X \cup \{\alpha\} \subseteq L$ is finite, then, by (ref) and $(\Delta_1)$,

$$\alpha \in C(X) \textit{ iff } \mathcal{A}_{con}(X_\alpha) = \textit{true}.$$

As far as the notion of consistency is concerned, the logics $\mathcal{P}$ and $\mathcal{P}_1$ are indistinguishable and, hence, we may consider $\mathcal{P}_1$ to be a refutationally equivalent formalization of $\mathcal{P}$. Therefore, one of the main tasks that we are faced with when searching for a refutational theorem proving method for a non-classical logic $\mathcal{P}$ is to *find a refutationally equivalent formalization $\mathcal{P}_1$ of $\mathcal{P}$ which can provide us with a consistency checking algorithm with desired properties (e.g., best computational complexity, certain behavior for certain classes of formulas, etc.).*

In this book, we approach this problem from one particular direction; we are concerned with the construction and analysis of refutationally equivalent formalizations based on the resolution principle.

## 1.3 PROPOSITIONAL LOGICS - SYNTAX

The purpose of this section is to review the syntax of propositional logics, and to formulate a minimal set of syntactic conditions that we want every propositional logic discussed in this book to satisfy.

Let $\{p_0, p_1, p_2, \ldots\}$ be a countably infinite set of symbols called *propositional variables* and let $\{f_0, \ldots, f_k\}$ be a finite set of symbols called *logical connectives*. We assume that both sets are disjoint. With every logical connective $f$ we associate an integer $\tau(f) \geq 0$, the *arity* of $f$. A connective $f$ such that $\tau(f) = 0$ is called a *logical constant*. Let $L$ be the smallest set such that:

(f1) every propositional variable and every logical constant is in $L$;

(f2) if $f$ is a connective of arity $t > 0$ and if $\alpha_1, \ldots, \alpha_t \in L$, then $f(\alpha_1, \ldots, \alpha_t) \in L$.

From the logical point of view, $L$ is the set of *well-formed formulas*. From the algebraic perspective, the connectives $f_0, \ldots, f_k$ are operations on $L$ and, hence, $\mathcal{L} = \langle L, f_0, \ldots, f_k \rangle$ forms an absolutely free algebra (frequently called an *algebra of formulas*) freely generated by the propositional variables. To stress the importance of the algebraic interpretation of $L$, we call $\mathcal{L} = \langle L, f_0, \ldots, f_k \rangle$ a *propositional language* and we define a *propositional logic* to be a pair $\langle \mathcal{L}, C \rangle$ rather than $\langle L, C \rangle$, where $C$ is a consequence operation on $L$.

Let $\langle \mathcal{L}, C \rangle$ be a propositional logic. If $\alpha, \beta \in L$, and if $p_i$ is a propositional variable, then we shall write $\alpha(p_i/\beta)$ to denote the result of simultaneous replacement of every occurrence of $p_i$ by $\beta$. Clearly, if $p_i$ does not occur in $\alpha$, then $\alpha(p_i/\beta) = \alpha$. The meaning of $\alpha(p_0/\beta_0, \ldots, p_k/\beta_k)$ is self-explanatory. If $X \subseteq L$, then by $Var(X)$ we denote the set of all variables occurring in formulas of $X$. Since $\mathcal{L}$ is a free algebra, every function mapping the variables of $\mathcal{L}$ into an algebra similar to $\mathcal{L}$

can be extended to a homomorphism. In particular, every function $e$ that maps the variables of $\mathcal{L}$ into $L$, extends to the endomorphism $e'$ of $\mathcal{L}$ such that for every formula $\alpha(p_0, \ldots, p_k) \in L, e'(\alpha) = \alpha(p_0/e(p_0), \ldots, p_k/e(p_k))$ (we assume that $p_0, \ldots, p_k$ are the only variables occurring in $\alpha$). From the logical point of view $e'$ can be thought of as a substitution function. For this reason we call endomorphisms of $\mathcal{L}$ *logical substitutions*.

If $X \cup \{\alpha, \beta\} \subseteq L$, then we shall frequently write '$C(X, \alpha, \beta)$' instead of '$C(X \cup \{\alpha, \beta\})$'. We say that a propositional logic $\langle \mathcal{L}, C \rangle$ is:

- *structural* if for every $X \subseteq L$ and every logical substitution $e$, $\alpha \in C(X)$ implies $e(\alpha) \in C(e(X))$;
- *compact* if for every $X \cup \{\alpha\} \subseteq L$, $\alpha \in C(X)$ implies $\alpha \in C(X_f)$, for some finite $X_f \subseteq X$;
- *disjunctive* if there is a binary connective $\vee$ (usually written as an infix operation) such that for every $X \cup \{\alpha, \beta\} \subseteq L$, $C(X, \alpha \vee \beta) = C(X, \alpha) \cap C(X, \beta)$.

Let $\langle \mathcal{L}, C \rangle$ be a propositional logic. Among the congruence relations of $\mathcal{L}$ there are two relations which will play an important role in our study of resolution proof systems. These congruences, denoted by $\Theta_C$ and $\Theta_C^R$, are defined as follows. For every $\alpha, \beta \in L$,

$$\alpha \, \Theta_C \, \beta \text{ iff for every } \gamma(p) \in L, C(\gamma(p/\alpha)) = C(\gamma(p/\beta));$$

$$\alpha \, \Theta_C^R \, \beta \text{ iff } C(X, \gamma(p/\alpha)) = L \Leftrightarrow C(X, \gamma(p/\beta)) = L, \text{ for every } X \cup \{\gamma(p)\} \subseteq L.$$

The first congruence $\Theta_C$ is the relation of inferential equivalence. Two formulas $\alpha$ and $\beta$ are $\Theta_C$-congruent if and only if they are indistinguishable as far as the process of deduction is concerned. On the other hand, $\alpha$ and $\beta$ are $\Theta_C^R$-congruent if and only if the replacement of $\alpha$ by $\beta$ (or vice versa) does not affect the consistency status of any context in which such a replacement has been performed.

Compact and structural logics are often called *standard* (cf. [79]). From now on, we shall concentrate exclusively on standard logics $\mathcal{P} = \langle \mathcal{L}, C \rangle$ which satisfy two additional properties:

- $C(\emptyset) \neq L$ and $C(Y) = L$, for some finite $Y \subseteq L$;
- $\mathcal{L}$ does not have $\Theta_C^R$-congruent constant symbols.

Both of these conditions are introduced with an eye on the existence and efficiency of refutational theorem proving methods. If $\mathcal{P}$ does not satisfy the first of these conditions, then refutational methods for $\mathcal{P}$ are either trivial (if $C(\emptyset) = L$, then, by (c2), every set of formulas is inconsistent in this logic) or are not available at all (indeed, if for every finite $Y \subseteq L, C(Y) \neq L$, then there are no finite sets of formulas that can be refuted). If the second condition is violated, i.e., if $\mathcal{L}$ has two different but $\Theta_C^R$-congruent constants, say $a_0$ and $a_1$, then one of these constants can simply be removed from $\mathcal{L}$. Indeed, if $X$ is an arbitrary set of formulas and if $X_1 \subseteq L$ is obtained by renaming $a_0$ as $a_1$ in every formula of $X$, then $X_1$ inherits the consistency status from $X$, i.e., $X_1$ is $C$-consistent if and only if $X$ is $C$-consistent.

To summarize, unless stated otherwise, the term 'propositional logic' means a standard logic satisfying the above two conditions. Examples of propositional logics can be found in Example 1.6 at the end of the next section, and in Example 4.9 of Section 4.2.

PROPOSITION 1.4: *Let $\langle \mathcal{L}, C\rangle$ be a propositional logic and let $X \subseteq L$ be finite. Then, $C(X) = L$ iff for every logical substitution $e$, $C(e(X)) = L$.*

*Proof:* If $C(X) \neq L$, then, $C(e(X)) \neq L$, where $e$ is the identity substitution. Let us suppose that $C(X) = L$ and let $e$ be any substitution. To show that $C(e(X)) = L$, we prove that every formula $\beta$ is in $C(e(X))$. To this end, let $\beta \in L$, let $p_i$ be a variable that does not belong to $Var(X)$, and let $e_\beta$ be the substitution such that: $e_\beta(p_i) = \beta$ and $e_\beta(p_j) = e(p_j)$, if $p_i \neq p_j$. Then $e_\beta(X) = e(X)$. Since $p_i \in C(X), \beta = e_\beta(p_i) \in C(e(X))$, by the structurality of the logic. ∎

PROPOSITION 1.5: *Let $\langle \mathcal{L}, C\rangle$ be a propositional logic. Then:*

(i) $\Theta_C \subseteq \Theta_C^R$;

(ii) *for every $\alpha, \beta \in L$, if $\alpha\ \Theta_C\ \beta$, then for every logical substitution $e$, $e(\alpha)\ \Theta_C\ e(\beta)$;*

(iii) *for every $\alpha, \beta \in L$, $\alpha\ \Theta_C^R\ \beta$ implies $e(\alpha)\ \Theta_C^R\ e(\beta)$, for every logical substitution $e$ which is a bijection from $Var(L)$ onto $Var(L)$.*

*Proof:* To show (i), let $\alpha\Theta_C\beta$ and let $X \cup \{\gamma(p)\} \subseteq L$. Then, $C(X, \gamma(p/\alpha)) = C(X, \gamma(p/\beta))$, which gives us $\alpha\Theta_C^R\beta$. To prove (ii), let us suppose that for some $\alpha, \beta \in L$, and some substitution $e, e(\alpha)$ and $e(\beta)$ are not $\Theta_C$-congruent. This means that for some $\gamma(p) \in L$, say, $\gamma(p/e(\alpha)) \notin C(\gamma(p/e(\beta)))$. Let $X = Var(\{\alpha, \beta, e(\alpha), e(\beta)\})$ and let $\gamma'$ be a formula obtained from $\gamma$ by renaming variables of $\gamma$ so that $Var(\gamma') \cap X = \emptyset$. Furthermore, let $e'$ be a substitution such that

$e'(\gamma') = \gamma$ and $e'(p_i) = e(p_i)$, for every $p_i \in X$. We claim that these assumptions imply

(a) $\gamma'(p_j/\alpha) \notin C(\gamma'(p_j/\beta))$,

and, hence, $\alpha$ and $\beta$ cannot be $\Theta_C$-congruent. To show (a), let us note that if $\gamma'(p_j/\alpha)$ were in $C(\gamma'(p_j/\beta))$, then, by structurality, $\gamma(p/e(\alpha)) = e'(\gamma'(p_j/\alpha)) \in C(e'(\gamma'(p_j/\beta))) = C(\gamma(p/e(\beta)))$, which is impossible. Finally, (iii) follows from Proposition 1.4 and the fact that $e^{-1}$ is a variable renaming substitution. ∎

## 1.4 PROPOSITIONAL LOGICS - SEMANTICS

This section, which describes the semantics of propositional logics, can be quickly passed over by anyone familiar with the theory of logical matrices.

Let $\mathcal{L} = \langle L, f_0, \ldots, f_k \rangle$ be an arbitrary propositional language. Every structural propositional logic on $\mathcal{L}$ can be semantically defined by a class of models called logical matrices. A *logical matrix* for $\mathcal{L}$ is a pair $M = \langle \mathcal{A}, d \rangle$. The first element, $\mathcal{A}$, of $M$ is an *algebra* $\mathcal{A} = \langle A, F_0, \ldots, F_k \rangle$ *of truth-values*, where $A$ is the set of truth-values and the operations $F_0, \ldots, F_k$ serve as the interpretations of the connectives $f_0, \ldots, f_k$, respectively. We assume that $\mathcal{L}$ and $\mathcal{A}$, as algebras, are of the same similarity type. The role of $\mathcal{A}$ is to provide the interpretation of logical connectives and to define the space of truth-values – the possible meanings of formulas of $\mathcal{L}$. The second element, $d$, of $M$ is a subset of $A$ and its elements are called the *designated truth-values* of $M$.

Since $\mathcal{L}$ is an absolutely free algebra, every function mapping the set of variables of $\mathcal{L}$ into the set of truth-values of $M$ can be extended to a function (to be precise, to a homomorphism of $\mathcal{L}$ into $\mathcal{A}$) that assigns a truth-value to every formula of $\mathcal{L}$. For this reason we call such homomorphisms *valuations* of $\mathcal{L}$ into $M$ and we denote the set of all such valuations by $Hom(\mathcal{L}, M)$.

With every logical matrix $M = \langle \mathcal{A}, d \rangle$ we associate the consequence operation $C_M$ defined in the following way: for every $X \cup \{\alpha\} \subseteq L$,

$(C_M)$ $\quad \alpha \in C_M(X)$ *iff* for every valuation $h$, $h(X) \subseteq d$ implies $h(\alpha) \in d$.

If $\mathcal{K}$ is a set of matrices for $\mathcal{L}$, then we let $C_{\mathcal{K}}$ denote the consequence operation defined by

$$C_{\mathcal{K}}(X) = \bigcap\{C_M(X) : M \in \mathcal{K}\}, \text{ all } X \subseteq L.$$

$\langle \mathcal{L}, C_{\mathcal{K}} \rangle$ is always structural but not necessarily standard. Moreover, for every structural logic $\langle \mathcal{L}, C \rangle$ there exists a class $\mathcal{K}$ of matrices such that $C = C_{\mathcal{K}}$. Hence, matrix semantics provides us with handy tools for the analysis and manipulation of structural logics; Wójcicki's [79] is a compendium on sixty years of research in matrix semantics for propositional calculi.

A collection $\mathcal{K} = \{\langle \mathcal{A}, d \rangle : d \in \mathcal{D}\}$ of matrices for $\mathcal{L}$ with the same underlying algebra $\mathcal{A}$ can be conveniently represented as a single generalized matrix $G = \langle \mathcal{A}, \mathcal{D} \rangle$. The consequence operation $C_G$ associated with the matrix $G$ is simply $C_{\mathcal{K}}$. Clearly, for every $X \subseteq L$ and every formula $\alpha \in L$,

$\alpha \in C_G(X)$ *iff* for every valuation $h$ and every $d \in \mathcal{D}$, $h(X) \subseteq d$ implies $h(\alpha) \in d$.

Since a logical matrix $\langle \mathcal{A}, d \rangle$ is a special case of a generalized matrix (it is a generalized matrix with just one set of designated truth-values) we shall write '$\langle \mathcal{A}, \{d\} \rangle$' instead of '$\langle \mathcal{A}, d \rangle$' and, henceforth, use the term 'logical matrix' instead of 'generalized logical matrix'. A matrix $M = \langle \mathcal{A}, \mathcal{D} \rangle$ is said to be *proper* if for every $d \in \mathcal{D}$ and every valuation $h$ of $\mathcal{L}$ into $M$, $h(L) \not\subseteq d$. In other words, no valuation assigns designated truth-values to all well-formed formulas.

The logics we discuss in the following example will be extensively used in the forthcoming chapters.

EXAMPLE 1.6: Let $\mathcal{L}$ be the propositional language with one unary connective $\neg$ (*negation*) and three binary connectives: $\vee$ (*disjunction*), $\wedge$ (*conjunction*), and $\rightarrow$ (*implication*). For every $n \geq 2$, the $n$-valued Łukasiewicz matrix for $\mathcal{L}$ is the system $M_n = \langle \mathcal{A}_n, \{n-1\} \rangle$, where

$$\mathcal{A}_n = \langle \{0, 1, \ldots, n-1\}, \underline{\neg}, \underline{\vee}, \underline{\wedge}, \underline{\rightarrow} \rangle,$$

is the so-called $n$-valued Łukasiewicz algebra with the operations $\underline{\neg}$, $\underline{\vee}$, $\underline{\wedge}$, and $\underline{\rightarrow}$ defined as follows:

$$\underline{\neg}x = n - 1 - x, \qquad x \underline{\vee} y = max(x, y),$$
$$x \underline{\wedge} y = min(x, y), \qquad x \underline{\rightarrow} y = min(n-1, n-1-x+y).$$

$n-1$ is the only designated truth-value of $M_n$. The logic $\mathcal{P}_n = \langle \mathcal{L}, C_n \rangle$ defined by $M_n$ is called the $n$-valued Łukasiewicz logic. In the light of this definition, $\mathcal{P}_2$ is the *classical propositional logic* while $\mathcal{P}_3$ is Łukasiewicz's famous three-valued logic. Every logic $\mathcal{P}_n$ satisfies the refutational principle (ref) defined in Section 1.2. Indeed, for every $n > 1$, there exists a formula $\sim(p)$ such that for every valuation $h$ of $\mathcal{L}$ into $M_n$,

$$h(\sim(p)) = \begin{cases} 0, & \text{if } h(p) = n-1, \\ n-1, & \text{otherwise.} \end{cases}$$

Hence, for every set $X \cup \{\alpha\} \subseteq L$,

$$\alpha \in C_n(X) \textit{ iff } C_n(X, \sim(p/\alpha)) = L.$$

For $\mathcal{P}_3$, we may define $\sim(p)$ as $\neg p \vee ((p \to \neg p) \wedge (\neg p \to p))$. The existence of $\sim(p)$ is guaranteed by the following test due to McNaughton (cf. [35]).

PROPOSITION 1.7 McNaughton Test: *Let $n \geq 2, k \geq 0$, and let $f$ be a $k$-ary operation on $A_n = \{0, 1, \ldots, n-1\}$. The following conditions are equivalent:*

(i) *$f$ is definable in $\mathcal{L}$, i.e., there exists a formula $\alpha_f(p_1, \ldots, p_k) \in L$ such that for every valuation $h$ of $\mathcal{L}$ into $M_n$, $f(h(p_1), \ldots, h(p_k)) = h(\alpha_f)$;*

(ii) *for every $a_1, \ldots, a_k \in A_n$, the greatest common divisor of $a_1, \ldots, a_k$ and $n-1$ is a divisor of $f(a_1, \ldots, a_k)$.*

In our second example we discuss the class of $n$-valued Post logics. For every $n \geq 2$, the $n$-valued *Post matrix* $M_n^*$ is defined in the same way as $M_n$ with the following two modifications. The operations $\underline{\to}$ and $\underline{\neg}$ are now defined by:

$$x \underline{\to} y = \begin{cases} n-1, & \text{if } x \leq y, \\ y, & \text{otherwise,} \end{cases}$$

$$\underline{\neg}(x) = \begin{cases} x-1, & \text{if } x > 0, \\ n-1, & \text{otherwise.} \end{cases}$$

The $n$-valued *Post logic* is the calculus $\mathcal{P}_n^*$ defined by $M_n^*$. This presentation of Post logics departs slightly from the original definition given in [51]. Let us note that every truth-value of $M_n^*$ is definable in $\mathcal{P}_n^*$. Indeed, let $v_{n-1}$ be $p_0 \to p_0$ and for every $i < n-1$, let $v_i = \neg^{(n-1-i)} v_{n-1}$. (For every formula $\alpha$ and every integer $k \geq 0$, $\neg^k \alpha$ is the abbreviation defined as follows: $\neg^0 \alpha = \alpha$ and $\neg^{k+1} \alpha = \neg\neg^k \alpha$.) Then, for every valuation $h$ of $\mathcal{L}$ into $M_n^*$, $h(v_i) = i$.

The final remark concerning Post logics is that, similarly to Łukasiewicz calculi, they satisfy the refutational principle. Indeed, for every $X \cup \{\alpha\} \subseteq L$, the set $X_\alpha = X \cup \{\neg\alpha \vee \neg^2\alpha \vee \ldots \vee \neg^{n-1}\alpha\}$ satisfies (ref). ∎

If $M = \langle \mathcal{A}, \mathcal{D} \rangle$ is a matrix, then by $|\mathcal{A}|$ we denote the set of truth-values of $M$. Let $M_0 = \langle \mathcal{A}_0, \{d_i^0 : i \in I\} \rangle$ and $M_1 = \langle \mathcal{A}_1, \{d_i^1 : i \in I\} \rangle$ be two matrices for $\mathcal{L}$. A mapping $h : \mathcal{A}_0 \to \mathcal{A}_1$ is called a *matrix homomorphism* of $M_0$ into $M_1$ if $h$ is a homomorphism of $\mathcal{A}_0$ into $\mathcal{A}_1$ and, for every $i \in I$ and every $a \in |\mathcal{A}_0|$,

$$a \in d_i^0 \textit{ iff } h(a) \in d_i^1.$$

If, in addition, $h$ is '1-1' and 'onto', then $h$ is called an *isomorphism*. Let $h$ be a matrix homomorphism of $M_0$ into $M_1$. The homomorphic image $h(M_0)$ of $M_0$ is the matrix $\langle h(\mathcal{A}_0), \{h(d_i^0) : i \in I\} \rangle$, where $h(\mathcal{A}_0)$ denotes the homomorphic image of the algebra $\mathcal{A}_0$.

We say that $M_1$ is a *submatrix* of $M_0$, provided that $\mathcal{A}_1$ is a subalgebra of $\mathcal{A}_0$ and for every $i \in I$,

$$d_i^1 = d_i^0 \cap |\mathcal{A}_1|.$$

PROPOSITION 1.8 (Wójcicki, cf. [79]): *Let $M_0$ and $M_1$ be two matrices for $\mathcal{L}$. Then:*

(i) *If $h$ is a matrix homomorphism of $M_0$ into $M_1$, then the homomorphic image $h(M_0)$ is a submatrix of $M_1$.*

(ii) *If $M_1$ is a homomorphic image of $M_0$, then $C_{M_0} = C_{M_1}$.*

(iii) *If $M_1$ is a submatrix of $M_0$, then for every $X \subseteq L, C_{M_0}(X) \subseteq C_{M_1}(X)$.*

Let $M = \langle \mathcal{A}, \mathcal{D} \rangle$ be a matrix and let $\Theta$ be a congruence of the algebra $\mathcal{A}$. $\Theta$ is said to be a *congruence* of $M$ if for every $d \in \mathcal{D}$ and every $a, b \in |\mathcal{A}|, a\Theta b$ and $a \in d$ implies $b \in d$. If $\Theta$ is a congruence of $M$, then the *quotient* $M/\Theta$ of $M$ is the matrix

$$M/\Theta = \langle \mathcal{A}/\Theta, \{d/\Theta : d \in \mathcal{D}\} \rangle,$$

where $\mathcal{A}/\Theta$ is the quotient of $\mathcal{A}$ modulo $\Theta$ and $d/\Theta$ is the set of equivalence classes of $\Theta$ determined by the elements of $d$.

PROPOSITION 1.9 (Wójcicki, cf. [79]): *Let $M$ be a matrix for $\mathcal{L}$ and let $\Theta$ be a congruence of $M$. Then $M/\Theta$ is a matrix for $\mathcal{L}$ and $C_M = C_{M/\Theta}$.*

We call a matrix $M = \langle \mathcal{A}, \mathcal{D} \rangle$ *factorial* if and only if the relation $\{\langle a, a \rangle : a \in |\mathcal{A}|\}$ (i.e., the identity relation on $|\mathcal{A}|$) is its only congruence.

By $card(M)$ we denote the *cardinality of a matrix $M$*, i.e., the number of truth-values of $M$. The logics defined by single finite matrices (i.e., single matrices of finite cardinalities) are of special interest to us. These calculi, called *strongly finite logics*, in addition to being structural are also compact (cf. [79], Theorem 4.1.7). If $\mathcal{L}$ is a propositional language and $k \geq 0$, then by $\mathcal{L}^{(k)}$ we denote the *sublanguage* of $\mathcal{L}$ generated by the variables $p_0, \ldots, p_k$. The set $L^{(k)}$ of well-formed formulas of $\mathcal{L}^{(k)}$ consists of all formulas of $\mathcal{L}$ built by means of the logical connectives of $\mathcal{L}$ and the propositional variables $p_0, \ldots, p_k$. The following criterion of strong finiteness is due to Wójcicki (cf. [79]).

PROPOSITION 1.10: *Let $\mathcal{P} = \langle \mathcal{L}, C \rangle$ be a structural propositional logic. Then $\mathcal{P}$ is strongly finite iff for some $k \geq 0$, the following two conditions are satisfied:*

(i) *for every $X \cup \{\alpha\} \subseteq L, \alpha \in C(X)$ iff $e(\alpha) \in C(e(X))$, for every substitution $e$ that maps $L$ into $L^{(k)}$;*

(ii) *the congruence $\Theta_C$ divides $L^{(k)}$ into finitely many equivalence classes.*

PROPOSITION 1.11 (Citkin, Zygmunt): *There is an algorithm to determine whether or not two finite matrices for $\mathcal{L}$ define the same consequence operations.*

The proof of this proposition can be found in [81].

The last two operations on matrices which we want to mention in this section are the direct and ramified products of matrices. Let $M_i = \langle \mathcal{A}_i, \{d_i\} \rangle, i = 0, 1$, be two matrices for $\mathcal{L}$. The *direct product* $M_0 \times M_1$ and the *ramified product* $M_0 \otimes M_1$ of $M_0$ and $M_1$ are logical matrices for $\mathcal{L}$ having the same underlying algebra $\mathcal{A} = \mathcal{A}_0 \times \mathcal{A}_1$ (the direct product of the algebras $\mathcal{A}_0$ and $\mathcal{A}_1$). $M_0 \times M_1$ has only one set of designated truth-values, the Cartesian product $d_0 \times d_1$. $M_0 \otimes M_1$ has two such sets: $|\mathcal{A}_0| \times d_1$ and $d_0 \times |\mathcal{A}_1|$.

PROPOSITION 1.12 (Maduch, Wójcicki, cf. [79]): *Let $M_i = \langle \mathcal{A}_i, \{d_i\} \rangle, i = 0, 1$, be two proper matrices for $\mathcal{L}$. Moreover, Let $C_\times$ and $C_\otimes$ denote the consequence operations defined by the matrices $M_0 \times M_1$ and $M_0 \otimes M_1$, respectively. Then $M_0 \times M_1$ and $M_0 \otimes M_1$ are proper and for every $X \subseteq L$:*

(i) $C_{\otimes}(X) = C_{M_0}(X) \cap C_{M_1}(X)$;

(ii)

$$C_{\times}(X) = \begin{cases} L, & \text{if } C_{M_0}(X) = L \text{ or } C_{M_1}(X) = L, \\ C_{M_0}(X) \cap C_{M_1}(X), & \text{otherwise.} \end{cases}$$

The notions of the direct and ramified products, as well as Proposition 1.12, can be generalized, in an obvious way, to any indexed family $\{M_i : i \in I\}$ of matrices. We shall write '$\Pi\{M_i : i \in I\}$' to denote the direct product of $\{M_i : i \in I\}$.

## 1.5 SEMANTIC TREES

The notion of a semantic tree discussed in this section will be used in Chapters 3 and 5 as the main tool in the proofs of refutational completeness.

Let $M = \langle \mathcal{A}, \{d\} \rangle$ be a finite matrix for $\mathcal{L}$, let $X \subseteq L$ be finite, and let $f$ be a partial function from $Var(X)$ into $|\mathcal{A}|$. $f$ provides interpretations of some (possibly none) of the variables of $Var(X)$ in the matrix $M$. A variable $p \in Var(X)$ is said to be free (bound) in $\alpha$ if $f(p)$ is undefined (defined). Let $p_1, \ldots, p_n$ be an enumeration (without repetitions) of free variables of all the formulas of $X$. A *semantic tree* $T_X$ of $X$ with respect to $M$ and $f$ represents all possible assignments of logical values of $M$ to the variables $p_1, \ldots, p_n$. The root node of $T_X$ represents the assignment $f$ which leaves $p_1, \ldots, p_n$ uninterpreted. The unique path from the root of $T_X$ to a node $N_k$ at level $k$ determines the interpretation $h_{N_k}$ which extends $f$ by assigning truth values to the first $k$ variables $p_1, \ldots, p_k$ and which is undefined for the remaining variables. A node $N_k$ of $T_X$ is a *failure node* if and only if for some $\alpha \in X$, $h_{N_k}(\alpha)$ is defined and $h_{N_k}(\alpha) \notin d$. $N_k$ is a leaf node if it is a failure node or $k = n$. Non-leaf nodes have $card(M)$ successors. The children of a non-leaf node $N_k$ represent partial valuations which extend $h_{N_k}$ by assigning different logical values to $p_{k+1}$ and which are undefined for the remaining variables $p_{k+2}, \ldots, p_n$.

A node $N_k$ is an *inference node* if it is a non-failure node all of whose children are failure nodes. A semantic tree $T_X$ of $X$ is said to be *closed* provided that every leaf of $T_X$ is a failure node.

EXAMPLE 1.13: Let $M_2$ be as in Example 1.6. Let $X = \{p_0 \vee (\neg p_1 \wedge p_2)\}$ and let $f$ be the partial function such that $f(p_0) = 0$ and $f$ is undefined for the remaining variables. The semantic tree $T_X$ of $X$ is shown in Figure 1.1.

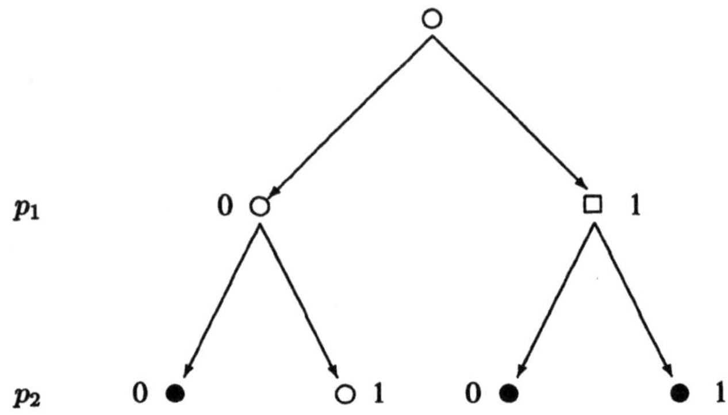


FIG. 1.1. The semantic tree of $\{p_0 \vee (\neg p_1 \wedge p_2)\}$ with respect to $M_2$ and $f$.

The ●, ○, and □ nodes represent failure, non-failure, and inference nodes, respectively. ∎

THEOREM 1.14: *Let $M = \langle \mathcal{A}, \{d\} \rangle$ be a finite proper matrix for $\mathcal{L}$ and let $X$ be a finite $C_M$-inconsistent set of formulas. Moreover, let $f$ be a partial function from $Var(X)$ into $|\mathcal{A}|$. Then every semantic tree of $X$ with respect to $M$ and $f$ is closed and is either one-element or has at least one inference node.*

*Proof:* Let $T_X$ be a semantic tree of $X$ with respect to $M$ and $f$, and let $p_1, \ldots, p_n$ be the free variables of formulas of $X$. Clearly, every leaf $N$ of $T_X$ at level $j < n$ is a failure node. Suppose that $N$ is a leaf of $T_X$ at level $n$. Let $h^* \in Hom(\mathcal{L}, M)$ be any extension of the partial valuation $h_N$. Since $X$ is $C_M$-inconsistent, there is a formula $\alpha \in X$ such that $h^*(\alpha) \notin d$. Hence $h_N(\alpha) = h^*(\alpha) \notin d$ and $N$ is a failure node.

Finally, let us note that if $T_X$ had more than one node, none of which were inference nodes, then there would exist a non-failure leaf. ∎

## 1.6 FIRST-ORDER LOGICS

A *first-order language* $\mathcal{L}$ is specified by four pairwise disjoint sets of symbols:

$X_L = \{x_0, x_1, \ldots\}$ - the set of *individual variables*,

$P_L$ - the set of *predicate symbols*,

$F_L$ - the set of *function symbols*,

$Con_L$ - the set of *logical connectives*,

and a function $\tau$ which assigns a natural number $\tau(f)$, the *arity* of $f$, to every symbol $f$ in $P_L \cup F_L \cup Con_L$. We assume that $P_L, F_L$, and $Con_L$ are finite and non-empty, that $F_L$ contains at least one *constant symbol* (i.e., a function symbol $c$ such that $\tau(c) = 0$), and that $X_L$ is countably infinite. We assume that for every $Q \in P_L$, $\tau(Q) > 0$, i.e., we identify 0-ary predicate symbols with logical constants.

We define the set of *terms* of $\mathcal{L}$ to be the smallest set $T_L$ satisfying the following conditions:

- every individual variable and every constant symbol is in $T_L$;
- if $f \in F_L$ is of arity $k > 0$, and if $t_1, \ldots, t_k \in T_L$, then $f(t_1, \ldots, t_k) \in T_L$.

We assume the usual recursive definition of the set $L$ of formulas of $\mathcal{L}$: $L$ is the smallest set which includes every logical constant and every *atomic formula* $Q(t_1, \ldots, t_k)$, where $Q \in P_L, \tau(Q) = k$, and $t_1, \ldots, t_k \in T_L$, and which is closed under the application of the connectives from $Con_L$. The terms '*closed term*', '*ground formula*', and '*subformula*' of a formula have their usual logical meanings.

Let us note that $\mathcal{L}$ is a quantifier-free language. We assume that every formula of $\mathcal{L}$ is implicitly universally quantified. This presentation of first-order languages reflects our desire to keep the theory of resolution proof systems separate from the part of logical theory that deals with the problem of quantifier elimination, the problem we shall briefly discuss at the end of Section 7.2.

Let $\mathcal{L}$ be an arbitrary first-order language, let $\mathcal{L}_p$ denote the propositional language which has the same logical connectives as $\mathcal{L}$, and let $M = \langle \mathcal{A}, \{d\} \rangle$ be a finite matrix for $\mathcal{L}_p$. The operations of $\mathcal{A}$ provide the interpretations of the connectives of $\mathcal{L}$. The semantics for $\mathcal{L}$ is defined by means of the notion of an interpretation of terms and formulas of $\mathcal{L}$ in a non-empty domain $U$ and the matrix $M$. An interpretation is a mapping $\pi$ which assigns to every function symbol $f \in F_L$ a mapping $\pi(f) : U^{\tau(f)} \to U$, and to every predicate symbol $Q \in P_L$, a mapping $\pi(Q) : U^{\tau(Q)} \to |\mathcal{A}|$. We call the pair $\langle U, \pi \rangle$ an *$M$-frame*. Any mapping $v : X_L \to U$ is called a *valuation* of individual variables. We extend $v$ to the valuation $v^*$ of $T_L \cup L$ into an $M$-frame $\langle U, \pi \rangle$ in the following way:

- $v^*(x_i) = v(x_i)$, if $x_i \in X_L$;
- $v^*(f(t_0, \ldots, t_k)) = \pi(f)(v^*(t_0), \ldots, v^*(t_k))$, if $f(t_0, \ldots, t_k)$ is a term or an atomic formula;
- $v^*(f(\alpha_0, \ldots, \alpha_k)) = f_M(v^*(\alpha_0), \ldots, v^*(\alpha_k))$, if $f(\alpha_0, \ldots, \alpha_k) \in L$ and $f \in Con_L$, where $f_M$ is the interpretation of $f$ in $M$.

An $M$-frame $\langle U, \pi \rangle$ is called an *$M$-valued model* of a formula $\alpha$ (or simply a model of $\alpha$) if for every valuation $v$ into $\langle U, \pi \rangle$, $v^*(\alpha) \in d$, i.e., $\alpha$ is 'true' in $M$ under all valuations. We say that $\langle U, \pi \rangle$ is a model of a set $X \subseteq L$, if $\langle U, \pi \rangle$ is a model of every formula in $X$.

Given a logical matrix $M = \langle \mathcal{A}, \mathcal{D} \rangle$ for $\mathcal{L}_p$, we define the consequence operation $C$ on $\mathcal{L}$ in the following way. For every $X \cup \{\alpha\} \subseteq L$,

> $\alpha \in C(X)$ *iff* for every $d \in \mathcal{D}$, every $\langle \mathcal{A}, \{d\} \rangle$-valued model of $X$ is a model of $\alpha$.

We call $\langle \mathcal{L}, C \rangle$ an $M$-valued *first-order logic* ($FFO$ logic). Similar formalizations of $FFO$ logics can be found in [20, 53, 76, 79].

EXAMPLE 1.15: Every logical system discussed in Example 1.6 has its $FFO$ counterpart, provided that the sets $P_L$ and $F_L$ are specified. Hence, classical first-order predicate logic is defined by the matrix $M_2$, and, in general, every Łukasiewicz matrix $M_n$ determines the class of $n$-valued Łukasiewicz $FFO$ logics. Similarly, every propositional Post logic has its $FFO$ counterparts. Further examples can be found in [53]. ∎

## 1.7 HERBRAND'S THEOREM

In this section we prove an analogue of Herbrand's Theorem (cf. [7]). Our theorem generalizes this classical result to $FFO$ logics. Herbrand's Theorem together with Lifting Lemma (cf. Lemma 7.6) will be used in Section 7.3 to prove the existence of resolution proof systems for $FFO$ logics.

Let $\mathcal{P} = \langle \mathcal{L}, C \rangle$ be an arbitrary but fixed $M$-valued $FFO$ logic, where $M = \langle \mathcal{A}, \mathcal{D} \rangle$ is a finite proper matrix. Let $\mathcal{L}_p$ denote the propositional language which has the same logical connectives as $\mathcal{P}$, and let $L_p$ denote the set of formulas of $\mathcal{L}_p$. We call $\mathcal{P}_p = \langle \mathcal{L}_p, C_M \rangle$ the *propositional logic associated* with $\mathcal{P}$. We assume that $\mathcal{P}_p$ has a finite inconsistent set, that $C_M(\emptyset) \neq L_p$, and that $\mathcal{P}$ has infinitely many ground atomic formulas.

For every non-empty set $A$ of constant symbols of $\mathcal{L}$, let $L^A$ denote the set of all formulas of $\mathcal{L}$ whose constant symbols are among the symbols of $A$. If $X \subseteq L$, then $H(X)$ denotes the set of all ground instances of formulas of $X$ and $H^A(X) = H(X) \cap L^A$.

Let $t$ be a one-one function mapping the set of all atomic formulas of $\mathcal{P}$ onto the set of propositional variables of $\mathcal{L}_p$. If $\alpha \in L$, then $t(\alpha)$ denotes the formula of $\mathcal{P}_p$ obtained from $\alpha$ by replacing every atomic subformula $\beta$ of $\alpha$ by $t(\beta)$. We call $t$ a *lifting function* for $\mathcal{P}$.

THEOREM 1.16 Herbrand's Theorem for $FFO$ Logics (O'Hearn, Stachniak): *Let $A$ be a nonempty set of constant symbols of $\mathcal{L}$. Then for every $X \subseteq L^A$ the following conditions are equivalent:*

(i) $C(X) = L$;

(ii) $C(Y) = L$, *for some finite subset $Y$ of $H^A(X)$;*

(iii) $C(H^A(X)) = L$;

(iv) $C_M(t(H^A(X))) = L_p$.

PROOF: (i)⇒(iv): Suppose that for some valuation $h \in Hom(\mathcal{L}_p, M)$, and some $d \in \mathcal{D}, h(t(H^A(X))) \subseteq d$. Let $\langle U, \pi \rangle$ be the $\langle \mathcal{A}, d \rangle$-frame consisting of the set $U$ of all closed terms of $L$ constructed from the constants of $A$, and of the mapping $\pi$ defined in the following way. If $a$ is a constant symbol, then put $\pi(a) = a$, if $a \in A$, and, for the remaining constants, let $\pi(a)$ be an arbitrary element of $U$. For every $k$-ary function symbol $f, k > 0, \pi(f)(t_1, \ldots, t_k) = f(t_1, \ldots, t_k)$. Finally, for every $k$-ary predicate $P, \pi(P)(t_1, \ldots, t_k) = h(t(P(t_1, \ldots, t_k)))$. Since for every valuation $v$ in

$\langle U, \pi\rangle$ and for every $\alpha \in X, v(\alpha) \in h(t(H^A(\{\alpha\}))) \subseteq d$, by Exercise 1.8, $C(X) \neq L$.

(iv)⇒(iii): Suppose that there exists an $\langle \mathcal{A}, d\rangle$-model $\langle U, \pi\rangle$ of $H^A(X)$. Let $v$ be any valuation in $\langle U, \pi\rangle$ and let $h \in Hom(\mathcal{L}_p, M)$ be selected so that for every atomic formula $A$ occurring in formulas of $H^A(X), h(t(A)) = v(A)$. Clearly, $h(t(H^A(X))) \subseteq d$ and, since $M$ is proper, $C_M(t(H^A(X))) \neq L_p$.

(iii)⇒(ii): Let $C(H^A(X)) = L$. If $C_M(t(H^A(X))) \neq L_p$, then there are $d \in \mathcal{D}$ and $h \in Hom(\mathcal{L}_p, M)$ such that $h(t(H^A(X))) \subseteq d$. Let $\langle U, \pi\rangle$ be an $\langle \mathcal{A}, d\rangle$-frame defined as in the proof of (i)⇒(iv). Then, for every valuation $v$ into $\langle U, \pi\rangle$ and every ground atomic formula $\alpha \in L^A, v(\alpha) = h(t(\alpha))$. Hence, $v(H^A(X)) = h(t(H^A(X))) \subseteq d$, which is impossible. So, $C_M(t(H^A(X))) = L_p$. By Exercise 1.4, $C_M(t(Y)) = L_p$, for some finite $Y \subseteq H^A(X)$. Using the same argument as in the proof of (iv)⇒(iii), we conclude that $C(Y) = L$.

(ii)⇒(i): Suppose that $C(H^A(X)) = L$. Let $\langle U, \pi\rangle$ be an $\langle \mathcal{A}, d\rangle$-frame where $d \in \mathcal{D}$. Moreover, let $\alpha \in H^A(X)$ and let $v$ be a valuation into $\langle U, \pi\rangle$ such that $v(\alpha) \notin d$. Then there is $\beta \in X$ and a substitution $e$ such that $e(\beta) = \alpha$. Let $v'$ be a valuation into $\langle U, \pi\rangle$ such that for every variable $x, v'(x) = v(e(x))$. So, $v'(\beta) = v(\alpha) \notin d$, which means that $X$ is inconsistent. ■

COROLLARY 1.17: *Let $M_0$ and $M_1$ be two finite proper matrices for $\mathcal{L}_p$ such that for every $X \subseteq L_p, C_{M_0}(X) = L_p$ iff $C_{M_1}(X) = L_p$. Let $\langle \mathcal{L}, C_0\rangle$ and $\langle \mathcal{L}, C_1\rangle$ be the FFO logics defined by $M_0$ and $M_1$, respectively. Then, for every $X \subseteq L, C_0(X) = L$ iff $C_1(X) = L$.*

*Proof:* Let $M_0$ and $M_1$ be as stated, let $X \subseteq L$, and let $t$ be a lifting function for $\langle \mathcal{L}, C_0\rangle$. By the hypothesis and by Herbrand's Theorem, $C_0(X) = L$ *iff* $C_{M_0}(t(H(X))) = L_p$ *iff* $C_{M_1}(t(H(X))) = L_p$ *iff* $C_1(X) = L$. ■

**EXERCISES**

1.1. Prove Theorem 1.2.

1.2. Let $\mathcal{L}$ be a propositional language with at least one connective of arity $\geq 2$ or at least two unary connectives. Show that there exists a consequence operation $C$ on $\mathcal{L}$ such that $\langle \mathcal{L}, C\rangle$ is decidable but every algorithm for constructing the deduced sets $X_\alpha$ is of non-polynomial time complexity.

1.3. Show that under our general assumptions concerning propositional logics the following holds. If $\mathcal{P}_0$ and $\mathcal{P}_1$ are two logics with the same finite inconsistent sets of formulas, then these logics have the same inconsistent sets of formulas.

1.4. (cf. [80]) Let $M$ be a finite matrix for $\mathcal{L}$ such that $C_M(\emptyset) \neq L$. Show that the following conditions are equivalent:

(i) $M$ is proper;

(ii) for every substitution $e$, $C_M(e(L)) = L$;

(iii) for every $X \subseteq L$, $C_M(X) = L$ implies $C_M(Y) = L$, for some finite subset $Y$ of $X$;

(iv) there exists a finite $C_M$-inconsistent set.

1.5. Let $M = \langle \mathcal{A}, \mathcal{D} \rangle$ be a matrix. Show that:

(i) $\bigcap\{\Theta_{C_{\langle \mathcal{A},d \rangle}} : d \in \mathcal{D}\} \subseteq \Theta_{C_M}$;

(ii) $\bigcap\{\Theta^R_{C_{\langle \mathcal{A},d \rangle}} : d \in \mathcal{D}\} \subseteq \Theta^R_{C_M}$.

1.6. (Cf. [42]) Let $\langle \mathcal{L}, C \rangle$ be a propositional logic. A congruence $\Theta$ of $\mathcal{L}$ is said to be *invariant* if and only if $\alpha\Theta\beta$ implies $e(\alpha)\Theta e(\beta)$, for every substitution $e$. Hence, by Proposition 1.5, $\Theta_C$ is invariant. Show that

$\Theta^R_C$ is invariant *iff* the quotient algebra $\mathcal{L}/\Theta^R_C$ is absolutely free.

1.7. Let $C_0$ and $C_1$ be two consequence operations on $\mathcal{L}$. Clearly, if $C_0$ and $C_1$ have the same inconsistent sets, then $\Theta^R_{C_0} = \Theta^R_{C_1}$. Show that $\Theta^R_{C_0} = \Theta^R_{C_1}$ does not always imply that $C_0$ and $C_1$ have the same inconsistent sets.

1.8. Let $\langle \mathcal{L}, C \rangle$ be an $M$-valued $FFO$ logic with infinitely many ground atomic formulas, where $M = \langle \mathcal{A}, \mathcal{D} \rangle$ is a finite proper matrix. Show that for every $X \subseteq L$, $C(X) = L$ *iff* for every $d \in \mathcal{D}$ and every $\langle \mathcal{A}, d \rangle$-frame $\Phi$, there is a valuation $v$ into $\Phi$ such that $v(X) \not\subseteq d$.

# 2

# PROPOSITIONAL RESOLUTION PROOF SYSTEMS

In this chapter we introduce two key notions of our theory of resolution proof systems: the notion of a propositional resolution proof system and the notion of a propositional resolution logic. Resolution proof systems studied in this book are deductive proof systems that are based on the non-clausal form of the resolution principle, i.e., formulas which are involved in the resolution based deductive process are not assumed to be in any special 'normal form'. Neither the relation between classical and non-classical automated proof methods nor the problem of 'clausal' versus 'non-clausal' proof procedures will be discussed in depth in this book. For the discussion on these and related topics the reader is referred to [19, 73, 75].

## 2.1 RESOLUTION PRINCIPLE

A refutational automated reasoning system can be viewed as a two module system. The first module accepts a query (i.e., a symbolic representation of a reasoning task) and converts it into its refutational representation $X_\alpha$ (i.e., into the deduced set). The second module represents the reasoning engine of the system and is used to determine the consistency status of $X_\alpha$. By (ref), a reasoning task terminates successfully as soon as $X_\alpha$ is refuted, i.e., proved inconsistent.

The theoretical principles underlying the construction of the first module, although far from being logically trivial, will not be further discussed in this book. Our main theoretical target is the study of refutational proof systems that can be used during the second phase of the deductive process, when the consistency status of the deduced set is to be determined.

The refutational proof systems studied in this book are called resolution proof systems, since their inference engine is based on the resolution principle. This principle, in its clausal form, was formulated by J. A. Robinson in [54] and used by him as the sole inference rule of the refutational counterpart of classical logic. Robinson's rule, restricted to the propositional case, operates on *clauses*, i.e., on finite sets of literals (a *literal* is a propositional variable or the negation of a propositional variable). This inference rule consists of all sequents of the form

$$R: \frac{A \cup \{p\}, B \cup \{\neg p\}}{A \cup B},$$

where $A \cup \{p\}$ and $B \cup \{\neg p\}$ are arbitrary clauses with complementary literals $p$ and $\neg p$. Since in classical logic a clause is interpreted as the disjunction of its literals, Robinson's rule $R$ can be thought of as an inference rule that operates on formulas.

The attractiveness of Robinson's rule stems from its syntactic simplicity and deductive power. Indeed, a simple iterative deductive process, at each step of which the resolution rule is applied to two clauses yielding a new clause, can be used to determine effectively the consistency status of every finite set of clauses. A finite set of clauses is inconsistent if and only if the empty clause (i.e., the empty set of literals) is generated during the deductive process. This, together with the refutational principle: for every finite set $X \cup \{\alpha\}$ of formulas of classical propositional logic $\mathcal{P}_2$,

> $X$ entails $\alpha$ in $\mathcal{P}_2$ *iff* the set of clauses representing $X \cup \{\neg\alpha\}$ is inconsistent in $\mathcal{P}_2$,

gives us a theorem proving algorithm for classical propositional logic (see [7, 54] for details).

Looking for ways of extending the clausal resolution method to non-classical logics we are faced with two difficulties. Robinson's rule operates on clauses, i.e., on formulas which have a special syntactic structure. Moreover, the formulation of the rule $R$ closely reflects the bivalent character of classical logic (we resolve only upon the complementary literals $p$ and $\neg p$). This makes it difficult (and frequently impossible) to adopt the notion of a clause and to generalize Robinson's rule in the context of non-classical reasoning while retaining the expressiveness and computational advantages exhibited by $R$ for classical logic.

The non-clausal variant of the principle (cf. [30, 39]) assumes the form

$$R^{\vee}_{\{F,T\}} : \frac{\alpha_0(p), \alpha_1(p)}{\alpha_0(p/F) \vee \alpha_1(p/T)},$$

where $\alpha_0, \alpha_1$ are arbitrary formulas that share propositional variable $p$, and $T$ and $F$ are formulas defining truth-values *truth* and *falsehood*, respectively. The formulas $F$ and $T$ 'witness' or 'verify' the consistency status of $\{\alpha_0(p), \alpha_1(p)\}$ or, as remarked in [30], they realize the 'case analysis on truth of a common subformula $p$': if $X \cup \{\alpha_0(p), \alpha_1(p)\}$ is consistent, then so is $X \cup \{\alpha_0(p), \alpha_1(p), \alpha_0(p/F) \vee \alpha_1(p/T)\}$. It can be easily verified that in classical propositional logic Robinson's rule $R$ is derivable from $R^{\vee}_{\{F,T\}}$.

To express the resolution principle in the context of non-classical logics we must take into account the fact that, in general, non-classical logics are neither disjunctive (i.e., the disjunction connective may not be definable) nor two-valued. Moreover, the resolution principle is an inference rule which means that $F$ and $T$, being formulas, cannot be simply replaced by some collection of truth-values. Our solution to the problem of generalization of the resolution rule to a non-classical logic $\mathcal{P}$ is based on the following interpretation of the rule.

(ver) *There exists a finite set $Ver$ of formulas of $\mathcal{P}$ with the help of which the consistency status of finite sets of formulas can be verified, i.e., if $X$ is a finite and consistent set of formulas of $\mathcal{P}$ and if $v_0, \ldots, v_n$ is an enumeration of the formulas of $Ver$, then for every choice $\alpha_0(p), \ldots, \alpha_n(p)$ of (not necessarily different) formulas of $X$, one of the sets $X \cup \{\alpha_0(p/v_0)\}, \ldots, X \cup \{\alpha_n(p/v_n)\}$ is also consistent.*

In (ver) we assume that $v_0, \ldots, v_n$ and the formulas of $X$ do not have any variables in common. We call (ver) the *verification* or *resolution principle*.

In order to express this principle as an inference rule, let $Ver = \{v_0, \ldots, v_n\}$ be a finite non-empty set of formulas of a propositional language $L$. The resolution rule determined by $Ver$ is the inference rule

(res) $$R_{Ver} : \frac{\alpha_0(p), \ldots, \alpha_n(p)}{\alpha_0(p/v_0), \ldots, \alpha_n(p/v_n)},$$

where $v_0, \ldots, v_n$ is an enumeration of $Ver$ and $\alpha_0(p), \ldots, \alpha_n(p)$ are arbitrary (not necessarily different) formulas with a common variable $p$. We assume that $p$ does not occur in formulas of $Ver$. Unless $L$ contains a disjunction connective (more accurately, a connective we intend to interpret as disjunction), $R_{Ver}$ is a multiple-conclusion rule.

If $L$ does contain a disjunction connective, say $\vee$, then the consequent of $R_{Ver}$, called the *resolvent* of $\alpha_0(p), \ldots, \alpha_n(p)$, can be written as $\alpha_0(p/v_0) \vee \ldots \vee \alpha_n(p/v_n)$, i.e., the resolution rule becomes the single-conclusion rule

$$R^{\vee}_{Ver}\colon \ \frac{\alpha_0(p), \ldots, \alpha_n(p)}{\alpha_0(p/v_0) \vee \ldots \vee \alpha_n(p/v_n)}.$$

(The superscript '$\vee$' used in $R^{\vee}_{Ver}$ indicates the single-conclusion, or disjunctive, variant of the resolution rule.) Clearly, if $L$ is the language of classical propositional logic and if we let $Ver = \{F, T\}$, then the resolution rule becomes the non-clausal rule $R^{\vee}_{\{F,T\}}$ discussed at the beginning of this section.

For the sake of our present discussion, let us assume that a disjunction connective is at our disposal. During the deductive process based on $R^{\vee}_{Ver}$ we attempt to show that a given deduced set is inconsistent. This process is carried out by expanding the deduced set by adding resolvents of formulas which are already in that set. The deductive process terminates successfully as soon as we are able to recognize an inconsistent subset of the deduced set. For instance, the deductive process based on $R^{\vee}_{\{F,T\}}$ terminates successfully, if a contradictory formula (i.e., a formula equivalent to $F$) is added to the deduced set. A simple mechanism for the identification of such formulas with $F$ is based on the successive simplifications of the deduced formulas using Boolean-like operations, as illustrated in the following example.

EXAMPLE 2.1: Let us consider classical propositional logic $\mathcal{P}_2$ (cf. Example 1.6). Let $T$ and $F$ stand for $p_0 \to p_0$ and $\neg T$, respectively. Moreover, let $\alpha = \neg p_2 \to \neg p_1$ and $X = \{p_1 \to p_2\}$. To show that $\alpha$ follows from $X$ in $\mathcal{P}_2$ we first form the deduced set $X_\alpha = X \cup \{\neg\alpha\}$, and then we subsequently add the following formulas to $X_\alpha$:

(1) $\neg(\neg p_2 \to \neg F) \vee (T \to p_2)$, (resolvent of $\neg\alpha$ and $p_1 \to p_2$)

(2) $\neg(\neg p_2 \to T) \vee (T \to p_2)$, (simplification of (1))

(3) $[\neg(\neg F \to T) \vee (T \to F)] \vee \neg(\neg T \to \neg p_1)$, (resolvent of (2) and $\neg\alpha$)

(4) $F \vee \neg(F \to \neg p_1)$, (simplification of (3))

(5) $[F \vee \neg(F \to \neg F)] \vee [F \vee \neg(F \to \neg T)]$, (resolvent of (4) and (4))

(6) $[F \vee \neg T] \vee [F \vee \neg T]$, (simplification of (5))

(7) $F$. (simplification of (6))

■

When a logical calculus does not have contradictory formulas, the termination of the deductive process based on the resolution rule requires a different solution than that described in Example 2.1. The solution employed in [43, 63, 64, 71] is to designate a certain family $\mathcal{F}$ of subsets of $Ver$ as *terminal*. During the deductive process we try to single out a subset of the deduced set that can be 'reduced' to, or identified with, one of the sets in $\mathcal{F}$, with the help of additional rules called the *simplification* or the *transformation* rules. The resolution rule, the family $\mathcal{F}$, and the transformation rules are the main building blocks of a deductive system called a *resolution proof system*.

EXAMPLE 2.2: Let the language $\mathcal{L}$ and the formulas $F$ and $T$ be as in Example 2.1. The resolution rule $R^{\vee}_{\{F,T\}}$ supplemented with the family $\mathcal{F}_2 = \{\{F\}, \{F, T\}\}$, and the transformation rules corresponding to the Boolean operations on the truth-values of classical propositional logic form a resolution proof system. In the refutation presented in Example 2.1, the transformation $\neg F \Rightarrow T$, has been applied to deduce (2) from (1) while $\neg F \Rightarrow T, \neg T \Rightarrow F, F \vee F \Rightarrow F, T \rightarrow T \Rightarrow T$, and $T \rightarrow F \Rightarrow F$ have been used to deduce (4) from (3). Since $F$ has been deduced in (7) and since $\{F\} \in \mathcal{F}_2$, the refutation of $X_\alpha$ terminates successfully. ∎

## 2.2 RESOLUTION PROOF SYSTEMS

Propositional resolution proof systems which we have informally introduced in the preceding section can be formally defined using algebraic structures called resolution algebras. Let $\mathcal{L} = \langle L, f_0, \ldots, f_k \rangle$ be an arbitrary but fixed propositional language. A *resolution algebra* on $\mathcal{L}$ is a pair

$$Rs = \langle \mathcal{V}, \mathcal{F} \rangle.$$

The first element of this pair, $\mathcal{V}$, is a finite algebra of the same similarity type as $\mathcal{L}$. The elements of this algebra are formulas of $\mathcal{L}$; we call them the *verifiers* of $Rs$. We make no special assumptions concerning the choice of verifiers with the exception that every logical constant of $\mathcal{L}$ (if they are present in $\mathcal{L}$) is a verifier of $Rs$. By $Ver_{Rs}$ we denote the set of all verifiers of $Rs$, i.e., $Ver_{Rs} = |\mathcal{V}|$. When $Rs$ is known from the context, we shall write '$Ver$' instead of '$Ver_{Rs}$'. The second element of $Rs$, $\mathcal{F}$, is a non-empty family of subsets of $Ver$.

With every resolution algebra $Rs$ we associate two types of inference rules: the resolution rule $R_{Ver}$ and the transformation rules. The definition of the resolution rule follows (res), i.e., $R_{Ver}$ is the set of all sequents of the form

$$\frac{\alpha_0(p), \ldots, \alpha_n(p)}{\alpha_0(p/v_0), \ldots, \alpha_n(p/v_n)},$$

where $\alpha_0, \ldots, \alpha_n$ are arbitrary formulas of $\mathcal{L}$ with a common variable $p$, and $v_0, \ldots, v_n$ is an enumeration of the set $Ver$ of verifiers. We assume that $p$ does not occur in the verifiers of $Rs$.

A *transformation rule* is any equality

$$f(w_1, \ldots, w_t) = w$$

that holds in $\mathcal{V}$. In this equality, $f$ is a $t$-ary operation of $\mathcal{V}$, $t > 0$, and $w, w_1, \ldots, w_t \in Ver$. From the algebraic point of view, the transformation rules define the operations of the algebra $\mathcal{V}$. However, if we identify $f$ with the connective of $\mathcal{L}$ it corresponds to, then the transformation rules become the rules of inference; we can read '$f(w_1, \ldots, w_t) = w$' as '$f(w_1, \ldots, w_t)$ can be rewritten as, or is reducible to $w$' and consider $w$ syntactically 'simpler' than $f(w_1, \ldots, w_t)$. While the resolution rule is the main inference rule of the resolution proof system, the transformation rules are simply rewrite rules used to simplify formulas generated during the deductive process. In the light of this interpretation, we shall frequently write '$f(w_1, \ldots, w_t) \Rightarrow w$' instead of '$f(w_1, \ldots, w_t) = w$'. Let us note that since the operations of $\mathcal{V}$ are total, the set of transformation rules is complete, i.e., every expression of the form $f(w_1, \ldots, w_t)$, where $f$ is a $t$-ary connective and $w_1, \ldots, w_t \in Ver$ is reducible to a verifier in $Ver$.

Let $Rs$ be a resolution algebra on $\mathcal{L}$. The resolution rule $R_{Ver}$ together with all the transformation rules of $Rs$ form the deductive basis of the *resolution proof system* based on $Rs$. Henceforth, we shall identify a resolution algebra with the resolution proof system it generates, and, when a confusion is likely to occur, we shall use the names of the connectives of $\mathcal{L}$ to denote the corresponding operations of $\mathcal{V}$. In addition, we reserve the term '$Ver_{Rs}$' (or '$Ver$', if $Rs$ is known from the context) to denote the set of verifiers of $Rs$.

## 2.3 DEDUCTIVE PROCESS

Let $Rs = \langle \mathcal{V}, \mathcal{F} \rangle$ be a resolution proof system on $\mathcal{L}$ having $n + 1$ verifiers. As we mentioned in the preceding section, the algebra $\mathcal{V}$ of $Rs$ defines the inference rules in terms of which the deductive process is carried out. Since $R_{Ver}$ is a multiple-conclusion rule, its application during the deductive process introduces branching, a case analysis of the consistency status of the set of formulas we want to refute. Hence, the deductive process based on the general form $R_{Ver}$ of the resolution rule assumes a tree structure. Its formal definition is as follows. Let $X$ be a finite set of formulas of $\mathcal{L}$. We shall write $X \overset{Rs}{\Rightarrow} \Box$ (and read '$X$ is refutable in $Rs$') if there is a finite tree $T$ (a *refutation tree* of $X$ in $Rs$) which satisfies the following conditions:

(t1) every node $N$ of $T$ is labelled with a set $X_N$ of formulas and has $0, 1$, or $n+1$ children; moreover, the label of the root of $T$ is $X$;

(t2) if $N_0$ is the only child of a node $N$ of $T$, then there is a formula $\alpha(p/w) \in X_N$ and a transformation rule $w \Rightarrow v$ such that $X_{N_0} = X_N \cup \{\alpha(p/v)\}$ (i.e., the child of $N_0$ is obtained from $N$ by the application of a transformation rule);

(t3) if a node $N$ has $n+1$ children, then there is a sequence $\alpha_0(p), \ldots, \alpha_n(p)$ of formulas of $X_N$ and an enumeration $v_0, \ldots, v_n$ of $Ver$ such that for every $i \leq n$, the $i$-th child of $N$ is labelled with $X_N \cup \{\alpha_i(p/v_i)\}$ (i.e., the children of $N$ are obtained by the application of the resolution rule of $Rs$);

(t4) the label of every leaf contains a set $Y \in \mathcal{F}$.

Conditions (t2) and (t3) describe the ways in which the inference rules of $Rs$ can be used during the deductive process. By (t2), if a transformation rule $w \Rightarrow v$ is applied to a formula $\alpha(p/w)$, then every occurrence of the subformula $w$ (the left-hand side of $w \Rightarrow v$) is rewritten as $v$ (the right-hand side of $w \Rightarrow v$) but not vice versa. While the first element of $Rs$ defines the inference rules in terms of which the deductive process is carried out, the role of the family $\mathcal{F}$ is to provide the termination conditions for such a process. By (t4), the deductive process terminates successfully if every branch of the refutation tree contains a terminal set from $\mathcal{F}$.

EXAMPLE 2.3: Let $\mathcal{L}$ be the language of classical propositional logic (cf. Example 1.6). The resolution proof system on $\mathcal{L}$ described in Example 2.2 uses a 'single-conclusion' form of the resolution rule. This type of the resolution rule has been discussed so far only informally and will be the subject of a detailed analysis in Section 3.3. The multiple-conclusion variant of this rule has the form

$$R_{\{F,T\}}\colon \frac{\alpha_0(p), \alpha_1(p)}{\alpha_0(p/v_0), \alpha_1(p/v_1)},$$

where $v_0, v_1$ is either the sequence $F, T$ or $T, F$, and the verifiers $F$ and $T$ are defined as in Example 2.1.

Let $Rs_2^* = \langle \mathcal{V}_2, \mathcal{F}_2 \rangle$ be the resolution proof system on $\mathcal{L}$ defined as follows. The algebra $\mathcal{V}_2$ consists of the set $\{F, T\}$ of verifiers and of the operations corresponding to the Boolean operations on the truth-values of classical logic. We accept $R_{\{F,T\}}$ as the resolution rule and $\mathcal{F}_2 = \{\{F\}, \{F,T\}\}$ as the family of terminal sets.

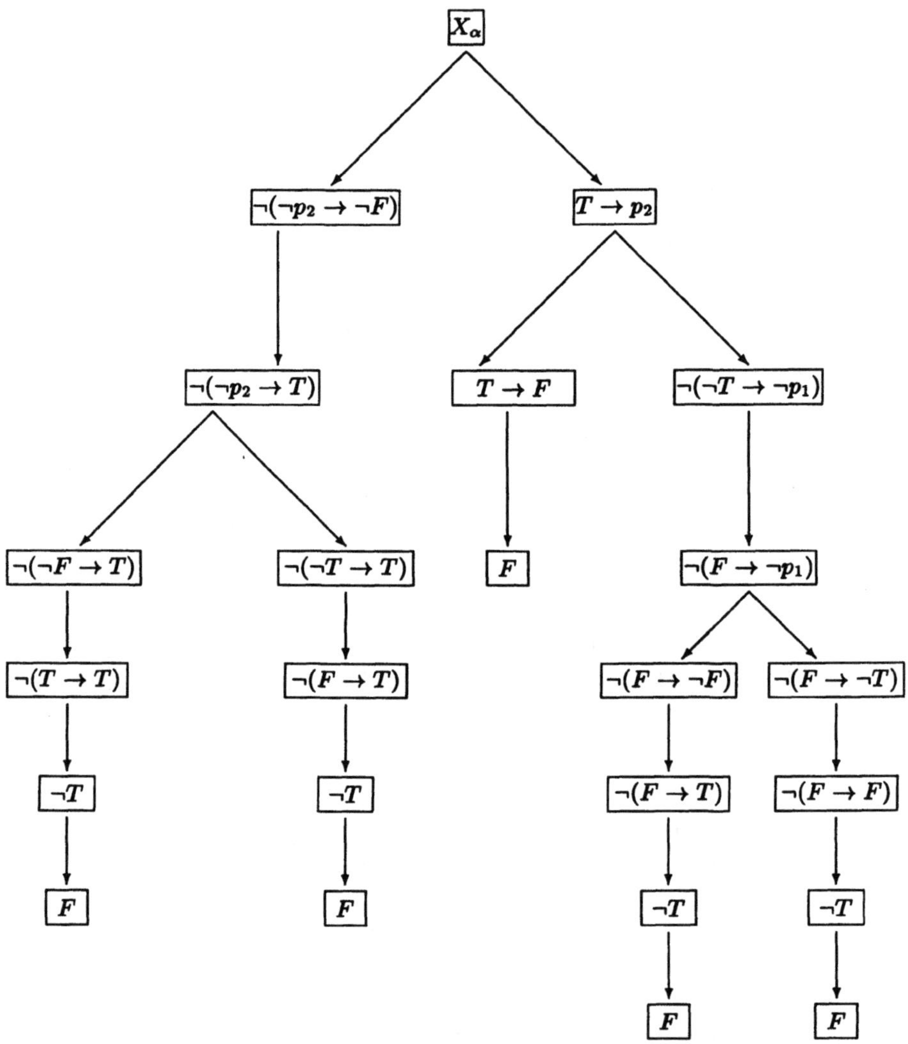


FIG. 2.1: Refutation tree of $\{p_1 \to p_2, \neg(\neg p_2 \to \neg p_1)\}$ in $Rs_2^*$.

To demonstrate $Rs_2^*$ in action, let us consider the set $X_\alpha = \{p_1 \rightarrow p_2, \neg(\neg p_2 \rightarrow \neg p_1)\}$. In Example 2.1, we have presented a refutation of $X_\alpha$ using the single-conclusion form of the resolution rule. Since we have to wait for this form of the resolution rule until Section 3.3, we need a refutation tree for $X_\alpha$. One of such trees is shown in Figure 2.1. The label of every node $N$ of this tree is the union of $X_\alpha$ and of the set of all the formulas we encounter while traversing the path from the root of the tree to $N$.

## 2.4 RESOLUTION LOGICS

We turn now to the task of linking propositional logics with resolution proof systems.

Let $\mathcal{P} = \langle \mathcal{L}, C\rangle$ be an arbitrary propositional logic and let $Rs = \langle \mathcal{V}, \mathcal{F}\rangle$ be a resolution proof system on $\mathcal{L}$. A set $X \subseteq L$ is said to be *clean* in $Rs$ (or simply *clean*, if $Rs$ is known from the context), if $Var(X) \cap Var(Ver) = \emptyset$. We call $Rs$ a *resolution counterpart* of $\mathcal{P}$ (and $\mathcal{P}$ a *resolution logic*) if:

(r1) for every finite and clean $X \subseteq L, C(X) = L$ *iff* $X \overset{Rs}{\Rightarrow} \Box$;

(r2) if $\phi \Rightarrow v$ is a transformation rule of $Rs$, then $\phi\ \Theta_C^R\ v$;

(r3*) for every $V \subseteq Ver, C(V) = L$ *iff* for some $Y \in \mathcal{F}, Y \subseteq V$.

(r1) expresses the *refutational completeness* of $Rs$ with respect to $\mathcal{P}$. The satisfaction of both (r1) and the refutational principle constitutes the foundation on which resolution based theorem proving procedures for $\mathcal{P}$ can be built; if $X \cup \{\alpha\} \subseteq L$, then, by (ref) and (r1), $\alpha \in C(X)$ if and only if $X_\alpha \overset{Rs}{\Rightarrow} \Box$.

The restriction of (r1) to clean sets only is not severe. What it asks for is to 'clean' a non-clean set $X$ before submitting it to a theorem prover, by renaming those variables occurring in formulas of $X$ that are in $Var(Ver)$. By Proposition 1.4, a clean version of $X$ is consistent if and only if the set $X$ is. Let us stress that if the restriction to clean sets is lifted, then the class of resolution counterparts of $\mathcal{P}$ may shrink as some of the resolution counterparts of $\mathcal{P}$ may refute some non-clean and $C$-consistent sets (cf. Example 3.16 in Chapter 3).

The refutational completeness principle (r1) provides also some justification for including logical constants (if they are present in $\mathcal{L}$) among the verifiers of $Rs$. For instance, if $c$ is a logical constant such that $\{c\}$ is inconsistent in $\mathcal{P}$, then, by (r1), $\{c\}$ is refutable in $Rs$. Clearly, this is only possible when $\{c\} \in \mathcal{F}$, i.e., when $c \in Ver_{Rs}$.

The intuition concerning (r2) is straightforward. The simplification of resolvents which are formed during the deductive process must preserve the consistency status of the deduced set.

Before we comment on (r3*), let us prove the following theorem.

THEOREM 2.4: *Let $Rs = \langle \mathcal{V}, \mathcal{F} \rangle$ be a resolution counterpart of a propositional logic $\mathcal{P}$ and let $Rs^* = \langle \mathcal{V}, \mathcal{F}^* \rangle$ be the resolution proof system, where $\mathcal{F}^*$ is the closure of $\mathcal{F}$ under supersets, i.e., $\mathcal{F}^* = \{V^* \subseteq Ver_{Rs} : for\ some\ V \in \mathcal{F}, V \subseteq V^*\}$. Then $Rs^*$ is a resolution counterpart of $\mathcal{P}$.*

PROOF: Let $\mathcal{P} = \langle \mathcal{L}, C \rangle, Rs$, and $Rs^*$ be as stated. Let $X \subseteq L$ be finite and clean in $Rs$. If $C(X) = L$, then there is a refutation tree $T$ of $X$ in $Rs$. Since $\mathcal{F} \subseteq \mathcal{F}^*, T$ is also a refutation tree of $X$ in $Rs^*$. Conversely, let $T$ be a refutation tree of $X$ in $Rs^*$ and let $N$ be any of its leaf nodes. The label of $N$ contains a set $V^* \in \mathcal{F}^*$. Since for some $V \in \mathcal{F}, V \subseteq V^*$, $N$ is a terminal node with respect to $Rs$ and, hence, $T$ is a refutation tree of $X$ in $Rs$. ∎

By (r3*), $\mathcal{F}$ is a family of inconsistent subsets of $Ver$. In view of Theorem 2.4 we may assume that $\mathcal{F}$ consists of all inconsistent subsets of $Ver$, and, hence, we may require that (r3*) be replaced with the simpler condition

(r3) for every $V \subseteq Ver, C(V) = L$ *iff* $V \in \mathcal{F}$.

EXAMPLE 2.5: The system $Rs_2^*$ defined in Example 2.3 is a resolution counterpart of classical propositional logic. For the proof, the reader is referred to Theorem 3.14 and Example 3.15 in the next chapter. ∎

LEMMA 2.6: *Let $Rs$ be a resolution proof system on $\mathcal{L}$ and let $X \subseteq L$ be finite and clean in $Rs$. If $X \overset{Rs}{\Rightarrow} \Box$, then there is a refutation tree $T$ of $X$ in $Rs$ such that the degree of every parent node $N$ of $T$ is $\geq$ to the degree of every child of $N$.*

*Proof:* Let $T'$ be a refutation tree of $X$ in $Rs$. Since we can only resolve upon variables that are not in $Var(Ver)$, and since the transformation rules of $Rs$ apply to formulas constructed from variables in $Var(Ver)$, we may postpone the application of transformation rules until all the applications of $R_{Ver}$ are done. In that way we can transform $T'$ into a refutation tree $T$ of $X$ in such a way that no application of a transformation rule precedes an application of the resolution rule. ∎

THEOREM 2.7: *Let $Rs$ be a resolution proof system on $\mathcal{L}$ and let $v_0, \ldots, v_n$ be a fixed enumeration of $Ver_{Rs}$. Moreover, let $X \subseteq L$ be finite and clean in $Rs$. If $X \overset{Rs}{\Rightarrow} \Box$, then there is a refutation tree of $X$ in $Rs$ such that every application of the resolution rule assumes this fixed enumeration of verifiers.*

*Proof:* Let $Rs$ and $v_0, \ldots, v_n$ be as stated. Let $X \subseteq L$ be finite, clean, and refutable in $Rs$. Finally, let $T$ be a refutation tree of $X$ in $Rs$. If $N$ is a node of $T$, then by $X_N$ we denote the label of $N$ and call $N$ a resolution node if it has more than one child. By Lemma 2.6 we can assume that in the construction of $T$, no application of a transformation rule precedes an application of the resolution rule. Let us assume

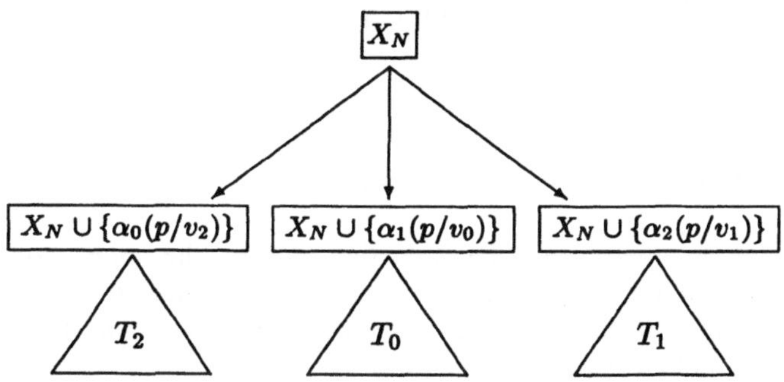


Before the rearrangement

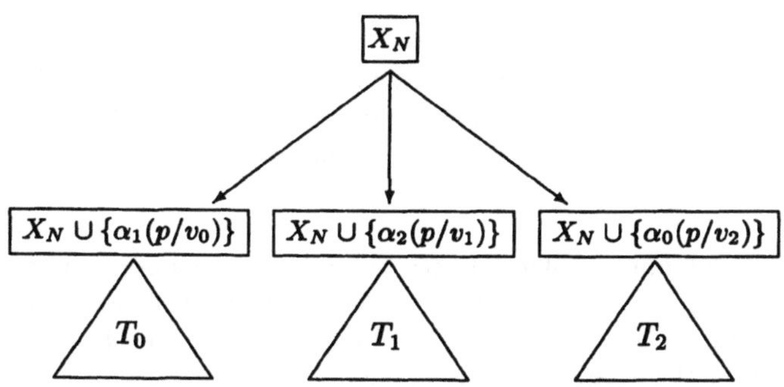


After the rearrangement

FIG. 2.2: The rearrangement of the children of a node $N$, for $n = 2$.

that $T$ has at least one resolution node and let $k$ be the last level of $T$ at which a resolution node appears. (We assume that the root of $T$ is at level 1.)

Next, let $T = T_1, \ldots, T_{k+1}$ be a sequence of refutation trees of $X$ constructed as follows. For every $1 \leq j \leq k, T_{j+1}$ is constructed from $T_j$ by processing every resolution node $N$ of $T_j$ at level $j$ in the following way. Let us assume that $N$ is a resolution node of $T_j$ at level $j$, and that for some $\alpha_0(p), \ldots, \alpha_n(p) \in X_N$, the $m$-th child of $N$ is labelled with $X_N \cup \{\alpha_m(p/v_{i_m})\}$, for some enumeration $v_{i_0}, \ldots, v_{i_n}$ of $Ver_{Rs}$.

Let $\alpha_{\rho(0)}, \ldots, \alpha_{\rho(n)}$ be the rearrangement of the children of $N$ (and, hence, of the subtrees of $T_j$ with the children of $N$ as their roots) so that now

(a) the $m$-th child of $N$ is labelled with $X_N \cup \{\alpha_{\rho(m)}(p/v_m)\}$

(see Figure 2.2). Clearly, $T_{j+1}$ is a refutation tree of $X$ in $Rs$. Moreover, every resolution node $N$ of $T_{j+1}$ at level $\leq j$ satisfies (a). Hence, $T_{k+1}$ is a refutation tree of $X$ that satisfies the conclusion of the theorem. ∎

Theorem 2.7 states that during the deductive process, the application of the resolution rule can be restricted to instances of this rule with an arbitrary but fixed enumeration of the verifiers of $Rs$. In other words, the resolution rule of $Rs$ can be defined as the set of all sequents of the form

$$\frac{\alpha_0(p), \ldots, \alpha_n(p)}{\alpha_0(p/v_0), \ldots, \alpha_n(p/v_n)},$$

where $\alpha_0, \ldots, \alpha_n$ are arbitrary formulas of $\mathcal{L}$ with a common variable $p$ not in $Var(Ver)$, and $v_0, \ldots, v_n$ is a fixed enumeration of $Ver$. Henceforth, we shall assume that the order of verifiers in the resolution rule of any resolution proof system is fixed.

Let $Rs$ be a resolution counterpart of a logic $\mathcal{P}$ and suppose that for some verifiers $v_0, v_1 \in Ver_{Rs}$, $v_0 \Theta^R_C v_1$. It turns out that if we eliminate one of these verifiers, say by renaming $v_0$ as $v_1$, then the resulting resolution proof system remains a counterpart of $\mathcal{P}$. The details are given in the following theorem.

THEOREM 2.8: *Let $Rs = \langle \mathcal{V}, \mathcal{F} \rangle$ be a resolution counterpart of a logic $\mathcal{P}$ and let $v_0, v_1 \in Ver_{Rs}$ be such that: $v_0$ is not a logical constant and $v_0 \ \Theta^R_C \ v_1$. Let $Rs' = \langle \mathcal{V}', \mathcal{F}' \rangle$ be the resolution proof system on $\mathcal{L}$ obtained from $Rs$ by renaming $v_0$ as $v_1$. That is:*

- $Ver_{Rs'} = Ver_{Rs} - \{v_0\}$;
- *the operations of* $\mathcal{V}'$ *are the restrictions of the operations of* $\mathcal{V}$ *to* $Ver_{Rs'}$ *except that if for some operation* $f$ *and some* $v_{i_0}, \ldots, v_{i_k} \in Ver_{Rs'}$, $f(v_{i_0}, \ldots, v_{i_k}) = v_0$ *in* $\mathcal{V}$, *then we put* $f(v_{i_0}, \ldots, v_{i_k}) = v_1$ *in* $\mathcal{V}'$;
- $\mathcal{F}'$ *is obtained from* $\mathcal{F}$ *by replacing* $v_0$ *by* $v_1$ *in every set of* $\mathcal{F}$.

*Then* $Rs'$ *is a resolution counterpart of* $\mathcal{P}$.

*Proof:* Let $Rs, Rs'$, and $\mathcal{P} = \langle \mathcal{L}, C \rangle$ be as stated. All we show is that $Rs'$ is refutationally complete with respect to $\mathcal{P}$. The rest of the proof is left to the reader as an exercise.

Let $X \subseteq L$ be finite and clean in $Rs'$. If $Var(X) = \emptyset$, then (r1) follows from the fact that every logical constant of $\mathcal{L}$ is a verifier of $Rs'$ and that $\mathcal{P}$ and $Rs'$ satisfy (r2) and (r3).

Next, let us assume that $Var(X) \neq \emptyset$. Without any loss of generality we can assume that $X$ is also clean in $Rs$. First, let us assume that $C(X) = L$ and let $T$ be a refutation tree of $X$ in $Rs$. By Lemma 2.6, we may assume that in the construction of $T$ no application of a transformation rule precedes an application of the resolution rule of $Rs$. Let $T_0$ be the tree obtained from $T$ by deleting all the nodes of $T$ which were created using transformation rules of $Rs$. Next, let $T_1$ be obtained from $T_0$ by pruning the leftmost branch of every node (starting from the root of $T_0$ and working down towards the leaf nodes). Clearly, the label of every leaf $N$ of $T_1$ contains a finite $C$-inconsistent subset $Y_N$ which is reducible to a set in $\mathcal{F}$ with the help of transformation rules of $Rs$. Since $Rs'$ and $\mathcal{P}$ satisfy (r2) and (r3), $Y_N$ is also reducible to a set in $\mathcal{F}'$ using the rules of $Rs'$. Hence, $T_1$ can be expanded to a refutation tree of $X$ in $Rs'$.

Conversely, let us suppose that $T'$ is a refutation tree of $X$ in $Rs'$. We want to prove that $C(X) = L$. As in the first part of the proof, we assume that in the construction of $T'$ no application of a transformation rule precedes an application of the resolution rule of $Rs'$. Let the tree $T^*$ be the result of removing all the nodes of $T'$ obtained with the help of transformation rules of $Rs'$. To simplify this part of the proof, we identify the nodes of $T^*$ with their labels. Moreover, we assume that $v_0, \ldots, v_n$ is the enumeration of verifiers used in the definition of the resolution rule of $Rs$.

Our first objective is to simulate the construction of $T^*$ by, roughly speaking, applying the instance

$$\frac{\alpha_1(p), \alpha_1(p), \ldots, \alpha_n(p)}{\alpha_1(p/v_0), \alpha_1(p/v_1), \ldots, \alpha_n(p/v_n)},$$

of the resolution rule of $Rs$ whenever an instance

$$\frac{\alpha_1(p), \ldots, \alpha_n(p)}{\alpha_1(p/v_1), \ldots, \alpha_n(p/v_n)},$$

of the resolution rule of $Rs'$ has been used. Let $k$ be the height of $T^*$ and let $T_1, \ldots, T_k$ be a sequence of trees satisfying the following properties. $T_1$ has only one node, $X$, and for every $1 \leq j < k$:

(a) $T_{j+1}$ is obtained from $T_j$ by expanding some of the leaves of $T_j$ by applying the resolution rule of $Rs$;

(b) there is a correspondence between the leaves of $T_j$ and some of the nodes of $T^*$ such that: if $N$ is a leaf of $T_j$ and $N'$ is the corresponding node of $T^*$, then there is a $1-1$ function $f$ mapping $N'$ onto $N$ such that for every $\beta \in N', f(\beta)$ is obtained from $\beta$ by replacing some (possibly none) of the occurrences of $v_1$ by $v_0$.

We begin the construction of the sequence in question by associating the only node of $T_1$ with the root of $T^*$. We also let $f$ be the identity function on $X$. Let us assume that for some $1 \leq j < k, T_1, \ldots, T_j$ have been constructed in accordance with (a) and (b). To construct $T_{j+1}$ we process every leaf $N$ of $T_j$ in the following way. Let $N'$ be the node of $T^*$ that corresponds to $N$ and let $f$ be the mapping satisfying (b) (see Figure 2.3). If $N'$ is a leaf of $T^*$, then $N$ is not processed. Otherwise, $N'$ has $n$ children, say $N' \cup \{\alpha_i(p/v_i)\}$, for some $\alpha_1, \ldots, \alpha_n \in N'$. Then every $f(\alpha_i)$ is in $N$ and we expand $N$ by resolving the sequence $f(\alpha_1), f(\alpha_1), \ldots, f(\alpha_n)$ upon $p$ in $Rs$. To satisfy (b), let the $i$-th child of $N$ correspond to the $(i-1)$-th child of $N'$, if $i > 1$, and, in addition, let the first child of $N$ correspond to the first child of $N'$. Moreover, for every child $N \cup \{f(\alpha_i)(p/v_j)\}$ of $N$ and its corresponding node $N' \cup \{\alpha_i(p/v_i)\}$ of $T^*$, the mapping satisfying (b) is constructed by extending $f$ so that $f(\alpha_i(p/v_i)) = f(\alpha_i)(p/v_j)$. This finishes the construction of $T_{j+1}$.

Now, let $N$ be a leaf of $T_k$, let $N'$ be its corresponding node in $T^*$, and let $f$ be the mapping satisfying (b). By the construction of $T^*$, there is a subset $K'$ of $N'$ which is reducible to some $V' \in \mathcal{F}'$. By (r3), $C(V') = L$ and, by (r2), $C(K') = L$.

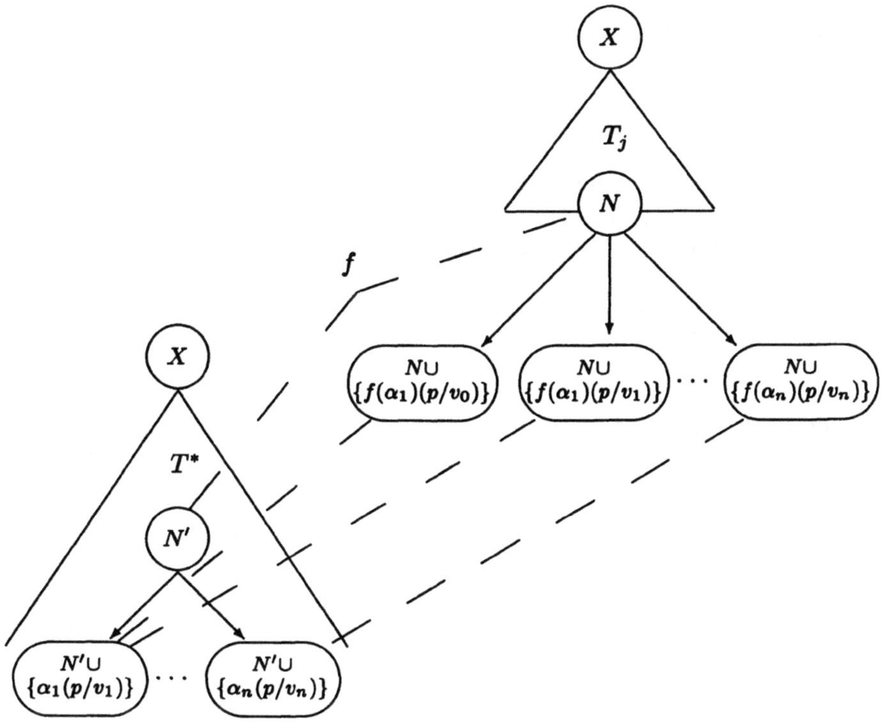


FIG. 2.3: The construction of $T_{j+1}$; the dashed lines join the corresponding nodes.

We claim that

(c) $f(K')$ can be reduced to a set $K \in \mathcal{F}$ using transformation rules of $Rs$.

From (a) and (c) it immediately follows that $T_k$ can be expanded to a refutation tree of $X$ in $Rs$ and, hence, that $C(X) = L$, as required. We begin the proof of (c) by noting that, in view of (b) and the fact that $X$ is clean in $Rs$, the formulas of $f(K')$ are built by means of verifiers of $Rs$ and the connectives of $\mathcal{L}$. Moreover, since $v_0 \Theta_C^R v_1$ and since $C(K') = L$, we also have $C(f(K')) = L$. Hence, by (r2), we can apply transformation rules of $Rs$ to reduce $f(K')$ to some $K \subseteq Ver_{Rs}$ such that $C(K) = L$. Now, (c) follows from (r3). ∎

From Theorem 2.8 we immediately obtain the following important facts:

COROLLARY 2.9: *Let $\mathcal{P}$ be a resolution logic. Then, there exists a resolution counterpart of $\mathcal{P}$ with no $\Theta_C^R$-congruent verifiers.*

*Proof:* Every resolution counterpart $Rs$ of $\mathcal{P}$ can be reduced to a resolution counterpart $Rs'$ of $\mathcal{P}$ with the required property by the successive removal of $\Theta_C^R$-congruent verifiers of $Rs$ in the way described in Theorem 2.8. ∎

COROLLARY 2.10: Let $\mathcal{P}$ be a resolution logic. Then, there exists a resolution counterpart $Rs$ of $\mathcal{P}$ such that:

(r0) for every $v_0, v_1 \in Ver$, if $v_1$ is obtained from $v_0$ by the application of a transformation rule, then $v_0 = v_1$.

In other words, (r0) says that the verifiers of $Rs$ are 'irreducible'.

*Proof:* Let $Rs$ be a resolution proof system satisfying Corollary 2.9. If $v_1$ is obtained from a verifier $v_0$ by the application of a transformation rule, then, by (r2), $v_0 \Theta_C^R v_1$. By Corollary 2.9, $v_0 = v_1$. ∎

From now on, in addition to (r1)–(r3), we shall require that resolution counterparts of logical calculi discussed in this book have no $\Theta_C^R$-congruent verifiers.

Theorem 2.7 and Corollaries 2.9 and 2.10 show that resolution proof systems can be tailored to specific requirements by controlling, to a certain extent, the number, the order, as well as syntactic structures of verifiers. The next two theorems, the proofs of which are left to the reader, can be interpreted in a similar manner.

THEOREM 2.11: *Let $Rs$ be a resolution counterpart of $\mathcal{P} = \langle \mathcal{L}, C \rangle$, let $v$ be a verifier of $Rs$ which is not a logical constant of $\mathcal{L}$, and let for some $w \in L, v \, \Theta_C^R \, w$. Let $Rs'$ be the resolution proof system on $\mathcal{L}$ obtained from $Rs$ by replacing $v$ by $w$. Then $Rs'$ is a resolution counterpart of $\mathcal{P}$.*

THEOREM 2.12: *Let $Rs$ be a resolution counterpart of a logic $\mathcal{P}$, and let $p$ and $q$ be variables such that $p \in Var(Ver_{Rs})$ while $q \notin Var(Ver_{Rs})$. Moreover, let $Rs'$ be the resolution proof system obtained from $Rs$ by renaming $p$ in all the verifiers of $Rs$ as $q$. Then $Rs'$ is a resolution counterpart of $\mathcal{P}$.*

COROLLARY 2.13: *Let Rs be a resolution counterpart of* $\mathcal{P} = \langle \mathcal{L}, C \rangle$, *and let* $k$ *be the cardinality of* $Var(Ver_{Rs})$. *Moreover, let* $W \subseteq L^{(k-1)}$ *be such that every logical constant of* $\mathcal{L}$ *is in* $W$ *and for every* $\alpha \in L^{(k-1)}$ *there is* $w \in W$ *such that* $\alpha\ \Theta_C^R\ w$. *Then there exists a resolution counterpart Rs' of* $\mathcal{P}$ *such that Rs and Rs' have the same number of verifiers and* $Ver_{Rs'} \subseteq W$.

*Proof:* Let $Rs, \mathcal{P}, k$, and $W$ be as stated. By Theorem 2.12, we can assume that $Ver_{Rs} \subseteq L^{(k-1)}$. By Theorem 2.11, we can replace the verifiers of $Rs$ with $\Theta_C^R$-congruent formulas from $W$. ∎

## 2.5 RESOLVING UPON SUBFORMULAS

The resolution rule, as defined in (res), operates on variables, i.e., it forms conclusions by replacing occurrences of variables with verifiers. It turns out that the deductive process can be based on a more general form of the resolution rule that allows resolution upon subformulas.

Let $Rs$ be a resolution proof system on $\mathcal{L}$ and let $R^s_{Ver}$ denote the inference rule defined by

$$R^s_{Ver}\colon \frac{\alpha_0(p/u),\ldots,\alpha_n(p/u)}{\alpha_0(p/v_0),\ldots,\alpha_n(p/v_n)},$$

where $v_0, \ldots, v_n$ is a fixed enumeration of $Ver_{Rs}$ and $\alpha_0(p/u), \ldots, \alpha_n(p/u)$ are arbitrary (not necessarily different) formulas with a common subformula $u$. We assume that $Var(u) \cap Var(Ver_{Rs}) = \emptyset$. Let the derivation relation $\overset{Rs}{\mapsto}$ be defined in the same way as the relation $\overset{Rs}{\Rightarrow}$ with the rule $R_{Ver}$ replaced by $R^s_{Ver}$.

THEOREM 2.14: *Let Rs be a resolution counterpart of a logic* $\langle \mathcal{L}, C \rangle$ *and let* $X \subseteq L$ *be clean in Rs. Then* $X \overset{Rs}{\Rightarrow} \Box$ *iff* $X \overset{Rs}{\mapsto} \Box$.

*Proof:* Clearly, $X \overset{Rs}{\Rightarrow} \Box$ implies $X \overset{Rs}{\mapsto} \Box$. Let us assume that $X \overset{Rs}{\mapsto} \Box$ and that $T$ is a refutation tree of $X$ defined in terms of $R^s_{Ver}$. Let $u_0, \ldots, u_k$ be the list of all subformulas of formulas of $X$ which, during the construction of $T$, have been replaced with verifiers using $R^s_{Ver}$. For every $i \leq k$, let $q_i$ be a variable that does not occur in $Var(X) \cup Var(Ver)$. Moreover, we assume that $q_i \neq q_j$, if $i \neq j$.

Next, for every $\alpha \in X$, let $A_\alpha$ be the set of all formulas that result from $\alpha$ by renaming all occurrences of some (or all) formulas $u_i$ by the corresponding $q_i$. Let us observe, that $T$ can be considered a refutation tree of $X' = \bigcup\{A_\alpha : \alpha \in X\}$ defined in terms of $R^s_{Ver}$ (we simply treat every formula $u_i$ as a propositional variable). Hence, $C(X') = L$. Let $e$ be a substitution such that $e(q_i) = u_i$, all $i \leq k$. By Proposition 1.4, $C(e(X')) = C(X) = L$, which, by (r1), gives us $X \overset{Rs}{\Rightarrow} \Box$. ∎

EXAMPLE 2.15: Let $Rs^*_2$ be the resolution counterpart of classical propositional logic defined in Example 2.3 and let $X = \{p_1 \to p_2, \neg(p_1 \to p_2)\}$. Clearly, $X$ is inconsistent in $\mathcal{P}_2$ and, hence, it is refutable in $Rs^*_2$. Using $R^s_{Ver}$ we can construct the following simple refutation tree of $X$:

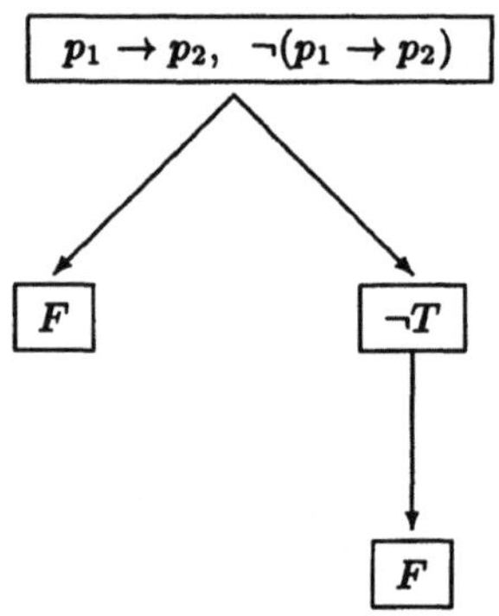


∎

## 2.6 STRONG RESOLUTION COUNTERPARTS

The notion of a resolution counterpart $Rs$ of a propositional logic $\mathcal{P}$ can be based on a stronger version of condition (r2). We may require that $Rs$ and $\mathcal{P}$ satisfy

(r2*) for every transformation rule $\phi \Rightarrow v$ of $Rs$, $\phi \Theta_C v$.

In other words, (r2*) requires any application of a transformation rule to preserve inferential, rather than refutational equivalence. We call $Rs$ *strong resolution counterpart* if every transformation rule of $Rs$ satisfies (r2*). By Proposition 1.5, $\Theta_C \subseteq \Theta^R_C$.

Hence, if a propositional logic $\mathcal{P}$ has a strong resolution counterpart, then $\mathcal{P}$ is a resolution logic. However, the converse of this fact is not true. In the next chapter we shall demonstrate that the class of resolution logics is essentially larger than the class of logics with strong resolution counterparts. For many resolution logics $\Theta_C^R = \Theta_C$ and, hence, (r2) and (r2*) are equivalent. Three classes of such logics are described in the following theorems.

THEOREM 2.16: *Let $\langle \mathcal{L}, C \rangle$ be a logic such that for some formula $\sim (p) \in L$ the following two conditions hold:*

(i) *for every $\beta \in L, C(\beta, \sim(p/\beta)) = L$;*

(ii) *for every $\alpha, \beta \in L, \alpha \notin C(\beta)$ implies $C(\beta, \sim(p/\alpha)) \neq L$.*

*Then $\Theta_C^R = \Theta_C$.*

*Proof:* Let $\langle \mathcal{L}, C \rangle$ and $\sim (p)$ be as stated. By Proposition 1.5, $\Theta_C \subseteq \Theta_C^R$. Let us suppose that some formulas $\alpha$ and $\beta$ are not $\Theta_C$-congruent. Then, for some $\gamma(q/\alpha)$, say, $\gamma(q/\alpha) \notin C(\gamma(q/\beta))$. By (ii), this means that $C(\gamma(q/\beta), \sim(\gamma(q/\alpha))) \neq L$. By (i), $C(\gamma(q/\beta), \sim(\gamma(q/\beta))) = L$. So, $\alpha$ and $\beta$ cannot be $\Theta_C^R$-congruent. ■

Every formula $\sim (p)$ satisfying Theorem 2.16(i) and (ii) defines a logical negation. Hence, the above theorem can be rephrased in the following way: for every resolution logic with negation the conditions (r2) and (r2*) are equivalent. The class of logics with negation includes, among other systems, every Łukasiewicz logic $\mathcal{P}_n$. Indeed, McNaughton Test (cf. Proposition 1.7) guarantees the existence of a formula $\sim (p)$ described in Theorem 2.16, for every logic $\mathcal{P}_n$.

The second class of logics for which the congruences $\Theta_C^R$ and $\Theta_C$ are identical is the class of the so-called definitionally complete logics. Let $\mathcal{A}$ be an algebra similar to $\mathcal{L}$ and let $|\mathcal{A}|$ denote the base set of $\mathcal{A}$, i.e., the set of elements of this algebra. We say that $\mathcal{A}$ is *definitionally complete* if and only if for every $n \geq 0$ and every function $f : |\mathcal{A}|^n \to |\mathcal{A}|$, there is a formula $\alpha_f(p_1, \ldots, p_n) \in L$ such that for every homomorphism $h$ of $\mathcal{L}$ into $\mathcal{A}$,

(d0) $f(h(p_1), \ldots, h(p_n)) = h(\alpha_f)$.

We say that a logic $\mathcal{P} = \langle \mathcal{L}, C \rangle$ is *definitionally complete* if and only if there exists a finite proper matrix $M = \langle \mathcal{A}, \mathcal{D} \rangle$ for $\mathcal{L}$ such that:

(d1) $\mathcal{A}$ is definitionally complete;

(d2) $M$ defines $\mathcal{P}$, i.e., $C_M = C$.

THEOREM 2.17: *If $\langle \mathcal{L}, C\rangle$ is a definitionally complete logic such that $C(\emptyset) \neq \emptyset$, then $\Theta_C^R = \Theta_C$.*

*Proof:* Let $\mathcal{P} = \langle \mathcal{L}, C\rangle$ be a definitionally complete logic with respect to a matrix $M = \langle \mathcal{A}, \mathcal{D}\rangle$. Let $\alpha, \beta$ be arbitrary formulas that are not $\Theta_C$-congruent. This means that for some $\gamma(p) \in L$, say, $\gamma(p/\alpha) \notin C(\gamma(p/\beta))$. Let $\alpha^*$ be $\gamma(p/\alpha)$ and $\beta^*$ be $\gamma(p/\beta)$. Since $C(\emptyset) \neq \emptyset, \bigcap \mathcal{D} \neq \emptyset$. Let $A = |\mathcal{A}|$. For every $b \in A$, let $f_b$ be the following function on $A$: for every $x, y \in A$,

$$f_b(x,y) = \begin{cases} a, & \text{if for some } d \in \mathcal{D}, x \notin d \text{ and } y \in d, \\ b, & \text{otherwise,} \end{cases}$$

where $a$ is a fixed member of $\bigcap \mathcal{D}$. Since $\mathcal{A}$ is definitionally complete, for every $b \in A$ there is a formula $\alpha_b(p_1, p_2)$ that defines $f_b$ in the sense of (d0). Let us verify that

(a) $C(\{\alpha_b(p_1/\alpha^*, p_2/\beta^*) : b \in A\}) \neq L$.

Since $\alpha^* \notin C(\beta^*)$, for some valuation $h$ and some $d \in \mathcal{D}$, $h(\beta^*) \in d$ while $h(\alpha^*) \notin d$. This means that for every $b \in A$, $h(\alpha_b(p_1/\alpha^*, p_2/\beta^*)) = f_b(h(\alpha^*), h(\beta^*)) = a \in d$, and (a) has been justified. Next, we claim that

(b) $C(\{\alpha_b(p_1/\beta^*, p_2/\beta^*) : b \in A\}) = L$.

To show (b), let us note that for every valuation $h'$ and every $d' \in \mathcal{D}$, $h'(\alpha_b(p_1/\beta^*, p_2/\beta^*)) = f_b(h'(\beta^*), h'(\beta^*)) = b$. Hence, $h'(\{\alpha_b(p_1/\beta^*, p_2/\beta^*) : b \in A\}) = A \not\subseteq d$, which justifies (b). By (a) and (b), there exists an enumeration $b_0, \ldots, b_k$ of $A$ and $j < k$, such that:

(c) $C(\alpha_{b_0}(p_1/\beta^*, p_2/\beta^*), \ldots, \alpha_{b_{j-1}}(p_1/\beta^*, p_2/\beta^*),$
$\alpha_{b_j}(p_1/\alpha^*, p_2/\beta^*), \ldots, \alpha_{b_k}(p_1/\alpha^*, p_2/\beta^*)) \neq L,$

(d) $C(\alpha_{b_0}(p_1/\beta^*, p_2/\beta^*), \ldots, \alpha_{b_j}(p_1/\beta^*, p_2/\beta^*),$
$\alpha_{b_{j+1}}(p_1/\alpha^*, p_2/\beta^*), \ldots, \alpha_{b_k}(p_1/\alpha^*, p_2/\beta^*)) = L.$

Let

$$X = \{\alpha_{b_0}(p_1/\beta^*, p_2/\beta^*), \ldots, \alpha_{b_{j-1}}(p_1/\beta^*, p_2/\beta^*), \alpha_{b_{j+1}}(p_1/\alpha^*, p_2/\beta^*), \ldots, \alpha_{b_k}(p_1/\alpha^*, p_2/\beta^*)\},$$

and let $\gamma^*(p) = \alpha_{b_j}(p_1/\gamma(p), p_2/\beta^*)$. [1] From (c) and (d) we immediately obtain $C(X, \gamma^*(p/\alpha)) \neq L$ while $C(X, \gamma^*(p/\beta)) = L$. This shows that $\alpha$ and $\beta$ are not $\Theta_C^R$-congruent. ■

We say that $\langle \mathcal{L}, C \rangle$ is an *implicational* logic if among the connectives of $\mathcal{L}$ there is a binary connective $\rightarrow$ that satisfies the following five conditions: for every $\alpha, \beta, \gamma \in L$,

(i1) $\alpha \rightarrow \alpha \in C(\emptyset)$,

(i2) $\beta \in C(\alpha, \alpha \rightarrow \beta)$,

(i3) $\alpha \rightarrow \gamma \in C(\alpha \rightarrow \beta, \beta \rightarrow \gamma)$,

(i4) $\beta \rightarrow \alpha \in C(\alpha)$,

(i5) for every variable $p$ of $\gamma$, $\gamma(p/\alpha) \rightarrow \gamma(p/\beta) \in C(\alpha \rightarrow \beta, \beta \rightarrow \alpha)$.

Of the logics we have discussed so far, classical propositional logic, Łukasiewicz logics $\mathcal{P}_n$, and Post logics $\mathcal{P}_n^*$ are implicational. Further examples can be found in [52].

THEOREM 2.18: *Let $\mathcal{P} = \langle \mathcal{L}, C \rangle$ be an implication logic defined by a matrix $M$ with one set of designated truth-values. If every element of $M$ is definable (in the sense of (d0)), then $\Theta_C^R = \Theta_C$.*

*Proof:* Let $\mathcal{P}$ and $M = \langle \mathcal{A}, \{d\} \rangle$ be as stated. Let $M^*$ be the quotient of $M$ modulo the greatest congruence $\Theta$ of $M$. Let us list some of the properties of $M^*$:

- $C = C_{M^*}$ (Proposition 1.9);
- the set of designated truth-values of $M^*$ consists of just one element which we denote by 1 (cf. [79], Lemma 3.5.2);

(a) for every truth-value $a$ of $M^*$, $a \rightarrow a = 1$ (cf. [79], Lemma 3.5.2);

(b) every truth-value of $M^*$ is definable.

---

[1] If $j = 0$, then (c), (d), and $X$ assume the forms:
$C(\alpha_{b_0}(p_1/\alpha^*, p_2/\beta^*), \ldots, \alpha_{b_k}(p_1/\alpha^*, p_2/\beta^*)) \neq L$,
$C(\alpha_{b_0}(p_1/\beta^*, p_2/\beta^*), \alpha_{b_1}(p_1/\alpha^*, p_2/\beta^*), \ldots, \alpha_{b_k}(p_1/\alpha^*, p_2/\beta^*)) = L$, and
$X = \{\alpha_{b_1}(p_1/\alpha^*, p_2/\beta^*), \ldots, \alpha_{b_k}(p_1/\alpha^*, p_2/\beta^*)\}$, respectively.

To prove (b), let $a$ be a truth-value of $M$, let $\alpha_a$ be a formula that defines $a$ in $M$, and let $[a]$ denote the congruence class of $\Theta$ determined by $a$. We claim that $\alpha_a$ also defines $[a]$ in $M^*$. To verify this claim, let $h$ be a valuation of $\mathcal{L}$ into $M^*$ and let $h^*$ be the homomorphism of $M$ into $M^*$ defined by $h^*(a) = [a]$, all $a \in |\mathcal{A}|$. Moreover, let $\underline{h}$ be any valuation of $\mathcal{L}$ into $M$ such that for every variable $p, \underline{h}(p) \in h(p)$ (see Figure 2.4). It is not difficult to verify that for every $\alpha \in L, h(\alpha) = h^*(\underline{h}(\alpha))$. This justifies the claim and (b).

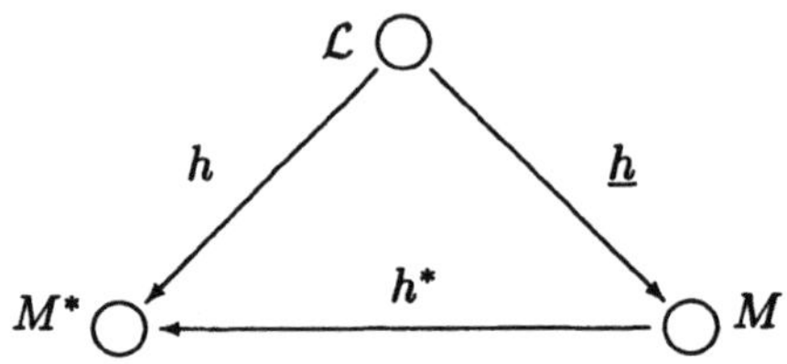


FIG. 2.4: The relationship between $h, h^*$, and $\underline{h}$.

Let us further observe that for every $\alpha, \beta \in L$,

(c) for every truth-value $[a]$ of $M^*$, $[a] \neq 1$ implies $C(\alpha, \alpha \to \alpha_a) = L$;

(d) $\alpha \notin C(\beta)$ implies $C(\beta, \alpha \to \alpha_a) \neq L$, for some $[a] \neq 1$.

To show (c), let $[a]$ be any truth-value of $M^*$. If $C(\alpha, \alpha \to \alpha_a) \neq L$, then for some valuation $h$ into $M^*$, $h(\{\alpha, \alpha \to \alpha_a\}) = \{1\}$. Since $\mathcal{P}$ is implicational, $\alpha_a \in C(\alpha, \alpha \to \alpha_a)$ and we must have $h(\alpha_a) = [a] = 1$. To prove (d), let us assume that there are $\alpha, \beta \in L$ such that for every $[a] \neq 1, C(\beta, \alpha \to \alpha_a) = L$. Let $h$ be a valuation of $\mathcal{L}$ into $M^*$ such that $h(\beta) = 1$. We claim that $h(\alpha) = 1$, and, hence, that $\alpha \in C(\beta)$. To prove the claim, let us note that for every $[a] \neq 1, h(\alpha \to \alpha_a) \neq 1$. By (a) this means that $h(\alpha) \neq [a]$, for every $[a] \neq 1$. Hence, $h(\alpha) = 1$. This concludes the proof of (d).

Now, we use (c) and (d) to prove our theorem. Let $\alpha$ and $\beta$ be arbitrary formulas that are not $\Theta_C$-congruent. This means that for some $\gamma(q) \in L$, say, $\gamma(q/\alpha) \notin C(\gamma(q/\beta))$. By (d), for some $[a] \neq 1$,

(e) $C(\gamma(q/\beta), \gamma(q/\alpha) \to \alpha_a) \neq L$.

By (c), for every $[a] \neq 1$,

(f) $C(\gamma(q/\alpha), \gamma(q/\alpha) \to \alpha_a) = L$.

If we let $X = \{\gamma(q/\alpha) \to \alpha_a\}$, then (e) and (f) become $C(X, \gamma(q/\alpha)) = L$ and $C(X, \gamma(q/\beta)) \neq L$. So, $\alpha$ and $\beta$ are not $\Theta_C^R$-congruent. ∎

The negation connective of the n-valued Post logic $\mathcal{P}_n^*$, $n > 2$, does not satisfy Theorem 2.16(ii). Hence, we cannot immediately conclude that for resolution counterparts of this logic, (r2) and (r2*) are equivalent. However, one can easily verify that $\mathcal{P}_n^*$ is implicational and, since every truth-value of the matrix $M_n^*$ is definable (see Example 1.6), we can use Theorem 2.18 to justify the equivalence of (r2) and (r2*) for these logics.

**EXERCISES**

2.1. Show that every resolution logic which satisfies the refutational principle (ref) defined in Section 1.2 is decidable.

2.2. Complete the proof of Theorem 2.8.

2.3. Let $Rs_0$ and $Rs_1$ be resolution counterparts of the same logic. Show that if $Rs_0$ and $Rs_1$ have the same sets of verifiers, then $Rs_0 = Rs_1$.

2.4. If $\mathcal{P} = \langle \mathcal{L}, C \rangle$ is a resolution logic, then by $\mathcal{RES}(\mathcal{P})$ we denote the class of all resolution proof systems $Rs$ on $\mathcal{L}$ that satisfy the following condition: for every finite $X \subseteq L$ clean in $Rs$,

$$C(X) = L \text{ implies } X \overset{Rs}{\Rightarrow} \Box.$$

Let $\mathcal{P}_i, i = 0, 1$, be two resolution logics. Show that these logics have the same inconsistent sets *iff* $\mathcal{RES}(\mathcal{P}_0) = \mathcal{RES}(\mathcal{P}_1)$.

2.5. Let $Rs_i = \langle \mathcal{V}_i, \mathcal{F}_i \rangle, i = 0, 1$, be two resolution proof systems on $\mathcal{L}$. A homomorphism $h$ of $\mathcal{V}_0$ into $\mathcal{V}_1$ is said to be an *r-homomorphism* of $Rs_0$ into $Rs_1$ if $\mathcal{F}_1 = \{h(f) : f \in \mathcal{F}_0\}$. $Rs_1$ is said to be a *subsystem* of $Rs_0$ if $\mathcal{V}_1$ is a subalgebra of $\mathcal{V}_0$ and $\mathcal{F}_1 = \{f \cap Ver_{Rs_1} : f \in \mathcal{F}_0\}$. Finally, we define the operators **H** and **S** on classes of resolution proof systems on $\mathcal{L}$ in the following way:

**H**($\mathcal{K}$): r-homomorphic images of systems of $\mathcal{K}$;

$\mathbf{S}(\mathcal{K})$: subsystems of resolution proof systems of $\mathcal{K}$.

Show that:

(a) $\mathbf{H}(\mathcal{RES}(\mathcal{P})) \subseteq \mathcal{RES}(\mathcal{P})$.

(b) $\mathbf{S}(\mathcal{RES}(\mathcal{P})) \subseteq \mathcal{RES}(\mathcal{P})$.

2.6. Let $Rs = \langle \mathcal{V}, \mathcal{F} \rangle$ be a strong resolution counterpart of a logic $\mathcal{P} = \langle \mathcal{L}, C \rangle$. Show that for every $\alpha, \beta \in L$, if $\alpha \ \Theta_C \ \beta$, then for every homomorphism $h$ of $\mathcal{L}$ into the algebra $\mathcal{V}$, $h(\alpha) = h(\beta)$.

2.7. Let $\mathcal{P}_n = \langle \mathcal{L}, C_n \rangle$ be the $n$-valued Łukasiewicz logic, let $M_n$ be the $n$-valued Łukasiewicz matrix, and let $\Theta_{M_n}$ be the congruence of $\mathcal{L}$ defined as follows: for every $\alpha, \beta \in L$,

$$\alpha \Theta_{M_n} \beta \quad \textit{iff} \quad \text{for every } h \in Hom(\mathcal{L}, M_n), h(\alpha) = h(\beta).$$

Show that $\Theta_{M_n} = \Theta_{C_n} = \Theta^R_{C_n}$.

2.8. (J. Słupecki) Show that an algebra $\mathcal{A}$ is definitionally complete *iff* the following functions on $\mathcal{A}$ are definable: all unary functions and at least one binary function that maps $|\mathcal{A}|$ onto $|\mathcal{A}|$.

# 3

# PROPOSITIONAL RESOLUTION LOGICS

In this chapter, we provide a characterization of the class of propositional resolution logics and delineate the distinct character of resolution logics with strong resolution counterparts. We also study resolution counterparts of disjunctive logics and show that in the presence of disjunction, the deductive process can be based on a single-conclusion variant of the resolution rule.

## 3.1 MATRICES INDUCED BY RESOLUTION PROOF SYSTEMS

Every resolution proof system $Rs$ induces, in a natural way, a finite logical matrix when the verifiers of $Rs$ are thought of as truth-values. This section brings a number of technical results which show that matrices induced by resolution proof systems are effective semantic tools for logical investigations of resolution logics.

We begin with the definition of the matrix induced by a resolution proof system. Let $Rs = \langle \mathcal{V}, \mathcal{F} \rangle$ be a resolution proof system on $\mathcal{L}$. By Theorem 2.4 we may assume that $\mathcal{F}$ is closed under supersets. Let $M_{Rs} = \langle \mathcal{V}, \mathcal{D} \rangle$ be the matrix for $\mathcal{L}$ defined as follows. We call a set $d \subseteq Ver_{Rs}$ *maximal consistent* in $Rs$ if $d \notin \mathcal{F}$ and for every $d' \subseteq Ver_{Rs}$, if $d \subseteq d'$ and $d' \notin \mathcal{F}$, then $d = d'$. The family $\mathcal{D}$ consists of all maximal consistent sets in $Rs$. We call $M_{Rs}$ the *matrix induced* by $Rs$; we call $\mathcal{P}_{Rs} = \langle \mathcal{L}, C_{M_{Rs}} \rangle$ the *logic induced* by $Rs$.

It is not difficult to verify that the logic induced by the resolution proof system $Rs_2^*$ of Example 2.3 is the classical logic $\mathcal{P}_2$. Indeed, the matrix $M$ induced by $Rs_2^*$ and the matrix $M_2$ which defines $\mathcal{P}_2$ (cf. Example 1.6) are isomorphic. Hence, by

Proposition 1.8, $C_M = C_2$.

Throughout this section we shall write '$C_{Rs}$' instead of '$C_{M_{Rs}}$' to denote the consequence operation defined by the matrix $M_{Rs}$ induced by a resolution proof system $Rs$. We begin our study of matrices induced by resolution proof systems with the following technical lemma.

LEMMA 3.1: *Let Rs be a resolution counterpart of a logic* $\langle \mathcal{L}, C \rangle$, *and let* $M_{Rs} = \langle \mathcal{V}, \mathcal{D} \rangle$ *be the matrix induced by Rs. Then:*

(i) *for every* $d \in \mathcal{D}, C(d) \neq L$ *and* $d = C(d) \cap Ver$;

(ii) *for every finite* $X \subseteq L, C_{Rs}(X) \neq L$ *implies* $C(h(X)) \neq L$, *for some* $h \in Hom(\mathcal{L}, M_{Rs})$;

(iii) *for every* $d \in \mathcal{D}$ *and every* $h \in Hom(\mathcal{L}, M_{Rs}), h(L) \not\subseteq d$;

(iv) *for every substitution* $e, C_{Rs}(e(L)) = L$.

*Proof:* If $d \in \mathcal{D}$, then by (r3), $C(d) \neq L$. Since $C(C(d) \cap Ver) \neq L$, the inclusion $d \subseteq C(d) \cap Ver$ cannot be proper; otherwise $d$ would not be maximal consistent in $Rs$. This concludes the proof of (i) from which (ii) follows immediately.

To prove (iii), suppose that for some $h \in Hom(\mathcal{L}, M_{Rs})$ and some $d \in \mathcal{D}, h(L) \subseteq d$. Then by (i),

(a) $C(h(Ver)) \subseteq C(h(L)) \subseteq C(d) \neq L$.

Let $e_h$ be the substitution defined in the following way. If $p \in Var(Ver_{Rs})$, then $e_h(p) = h(p)$; otherwise $e_h(p) = p$. With the help of the transformation rules of $Rs$ we can reduce $e_h(Ver)$ to $h(Ver)$. So, by (r2) and (a),

(b) $C(e_h(Ver)) \neq L$.

However, by (r3), $C(Ver) = L$. So, by Proposition 1.4, $C(e_h(Ver)) = L$. This contradicts (b) and completes the proof of (iii).

To prove (iv), suppose that for some substitution $e, C_{Rs}(e(L)) \neq L$. Then for some $h \in Hom(\mathcal{L}, M_{Rs})$ and some $d \in \mathcal{D}, h(e(L)) \subseteq d$. However, $e \circ h \in Hom(\mathcal{L}, M_{Rs})$

and $e \circ h(L) \subseteq d$, which contradicts (iii). ∎

We call a resolution proof system $Rs$ *proper* if it satisfies Lemma 3.1(iii), i.e., no $C_{Rs}$-inconsistent set is satisfiable in $M_{Rs}$. As the next theorem shows, 'properness' can be determined effectively.

THEOREM 3.2: *There is an effective way to determine whether or not a resolution proof system is proper.*

*Proof:* Our proof follows the proof of Corollary 2 in [80]. Let $Rs = \langle \mathcal{V}, \mathcal{F} \rangle$ be a resolution proof system on $\mathcal{L}$ and let $M_{Rs} = \langle \mathcal{V}, \mathcal{D} \rangle$ be the matrix induced by $Rs$. We claim that $Rs$ is proper if and only if

(a) for every subalgebra $\mathcal{B}$ of $\mathcal{V}$ and every $d \in \mathcal{D}$, $|\mathcal{B}| \not\subseteq d$,

where $|\mathcal{B}|$ denotes the base set of $\mathcal{B}$. To prove the claim, suppose that for some subalgebra $\mathcal{B}$ of $\mathcal{V}$ and some $d \in \mathcal{D}, |\mathcal{B}| \subseteq d$. Let $h$ be any homomorphism form $\mathcal{L}$ into $\mathcal{B}$. Then $h(L) \subseteq |\mathcal{B}| \subseteq d$. Since $h$ is also a homomorphism of $\mathcal{L}$ into $M_{Rs}$, $Rs$ cannot be proper.

Conversely, if for some $h \in Hom(\mathcal{L}, M_{Rs})$ and some $d \in \mathcal{D}, h(L) \subseteq d$, then $h(\mathcal{L})$ is a subalgebra of $\mathcal{V}$ which violates (a). This proves the claim from which our theorem follows immediately. ∎

From now on, by 'resolution proof system' we shall always mean 'proper resolution proof system'. Perhaps the most important among the properties of the matrix $M_{Rs}$ induced by a resolution proof system $Rs$ is that $M_{Rs}$ semantically defines all the finite sets refutable in $Rs$. The proof of this property follows from the following two lemmas.

LEMMA 3.3: *Let $Rs$ be a resolution proof system on $\mathcal{L}$ and let $X \subseteq L$ be finite and clean in $Rs$. Then $C_{Rs}(X) \neq L$ implies $X \overset{Rs}{\not\Rightarrow} \square$.*

PROOF: Let $Rs$ and $X$ be as stated. We want to show that $X$ cannot be at the same time $C_{Rs}$-consistent and refutable in $Rs$. Suppose that it can, i.e., that there are $h \in Hom(\mathcal{L}, M_{Rs})$ and $d \in \mathcal{D}$, such that $h(X) \subseteq d$ and, at the same time, that there exists a refutation tree $T$ of $X$ in $Rs$. In the light of Lemma 2.6 we can assume that none of the transformation rules of $Rs$ have been applied in $T$ before any application of the resolution rule.

We claim that there is a leaf $N$ of $T$ such that

(a) $X_N \cap Ver \subseteq h(X)$, where $X_N$ is the label of $N$.

A leaf $N$ satisfying (a) can be selected as follows. Let us define a path $N_0, \ldots, N_k$ of $T$ in the following way. $N_0$ is the root of $T$. Suppose that $N_i$ has been selected. If $N_i$ has been expanded in $T$ by the application of the resolution rule to the sequence $\alpha_0(p), \ldots, \alpha_n(p)$ of formulas of $X_{N_i}$, then let $N_{i+1}$ be the child of $N_i$ labelled with $X_{N_i} \cup \{\alpha_j(p/h(p))\}$. If $N_i$ has been expanded by the application of a transformation rule, then $N_{i+1}$ is the child of $N_i$.

Let $N_j$ be the first node in the path to which a transformation rule has been applied. Each formula of $X_{N_j}$ that is reduced in the path to a verifier must be of the form $\beta(p_0/h(p_0), \ldots, p_s/h(p_s))$, for some $\beta(p_0, \ldots, p_s) \in X$. Since transformations in $Rs$ are defined by the operations of $\mathcal{V}$, the formula $\beta(p_0/h(p_0), \ldots, p_s/h(p_s))$ reduces to the verifier $h(\beta(p_0, \ldots, p_s))$. Hence, $X_{N_k} \cap Ver$ consists of formulas of the form $h(\beta(p_0, \ldots, p_s))$. This gives us $X_{N_k} \cap Ver \subseteq h(X)$ and, by letting $N = N_k$, our search for a leaf of $T$ that satisfies (a) has been finished.

From (a) we conclude that no subset of $X_N \cap Ver$ is in $\mathcal{F}$. This, of course, contradicts the assumption that $T$ is a refutation tree of $X$. ∎

LEMMA 3.4: *Let Rs and X be as in Lemma 3.3. Then $C_{Rs}(X) = L$ implies $X \overset{Rs}{\Rightarrow} \Box$.*

PROOF: Let $X \subseteq L$ be finite, $C_{Rs}$-inconsistent, and clean in $Rs$. For every $v \in Ver$, let $p_v$ be a new variable not in $Var(X \cup Ver)$ and let $f$ be a partial function defined by: $f(p_v) = v$, all $v \in Ver$. Finally, let $d_0, \ldots, d_m$ be an enumeration of the family $\mathcal{D}$ of all sets maximal consistent in $Rs$. Our objective is to construct a refutation tree of $X$ in $Rs$.

First, we construct a certain finite tree $T$ and label its every node $N$ with a set $X_N \subseteq L$. In addition, with every node $N$ of $T$ we associated a sequence $S_0^N, \ldots, S_m^N$ of trees such that

(a) every $S_j^N, j \leq m$, is a closed semantic tree of $X_N$ with respect to $f$ and the matrix $\langle \mathcal{V}, \{d_j\} \rangle$.

We construct $T$ level by level. The root $Rt$ of $T$ is labelled by $X$. In addition, for each $j \leq m$, let $S_j^{Rt}$ be any closed semantic tree of $X$ with respect to $f$ and $\langle \mathcal{V}, \{d_j\} \rangle$. By Theorem 1.14, such trees can always be selected since $Rs$ is proper and for every

$j \leq m, C_{\langle \mathcal{V},\{d_j\}\rangle}(X) = L$. (Note, that $C_{Rs}(X) = \bigcap\{C_{\langle \mathcal{V},\{d\}\rangle}(X) : d \in \mathcal{D}\}$.) Every $S_j^{Rt}$ is either a one-element tree or has at least one inference node.

Suppose that $T$ has been constructed until level $k \geq 1$ and that all the nodes of $T$ satisfy (a). With every node of $T$ at level $k$ we proceed as follows. Let $S_0^N, \ldots, S_m^N$ be the closed semantic trees associated with a node $N$ at level $k$. If all the trees $S_0^N, \ldots, S_m^N$ are one-element, then $N$ is not processed. If for some $j \leq m, S_j^N$ is not one-element, then $S_j^N$ must have at least one inference node, say $K$. Every child $K_i$ of $K$, $i \leq n$, falsifies some formula $\alpha_i \in X_N$. By simulating an application of $R_{Ver}$ to $\alpha_0(p_{k+1}), \ldots, \alpha_n(p_{k+1})$, we form $n+1$ children $N_i$ of $N$ labelled by $X_N \cup \{\alpha_i(p_{k+1}/p_{v_i})\}$. (The variable $p_{v_i}$ will be later replaced by the verifier $v_i$.) For each $i \leq n$ and $j \leq m$, the tree $S_j^{N_i}$ is obtained from $S_j^N$ by removing the children $K_0^*, \ldots, K_n^*$ of each inference node $K^*$ of $S_j^N$, if $K_0^*, \ldots, K_n^*$ falsify the formulas $\alpha_0(p_{k+1}), \ldots, \alpha_n(p_{k+1})$, respectively. Clearly, $S_j^{N_i}$ remains closed since either $S_j^{N_i} = S_j^N$ or $K^*$ falsifies $\alpha_i(p_{k+1}/p_{v_i})$. Moreover, there are fewer inference nodes in $S_0^{N_i}, \ldots, S_m^{N_i}$ than in $S_0^N, \ldots, S_m^N$, for each $i$.

After a finite number of similar steps, every leaf of $T$ will have a sequence of one-element semantic trees associated with it. This concludes the construction of $T$.

Any one-element closed semantic tree of a set $X_N$ with respect to $f$ and $\langle \mathcal{V}, \{d_j\}\rangle$, $j \leq m$, must falsify some formula $\alpha_j \in X_N$ without free variables. Hence, for every leaf $N$ of $T$ and every $j \leq m$, there is $\alpha_j \in X_N$ such that $f(\alpha_j)$ evaluates in $\mathcal{V}$ to a verifier that does not belong to $d_j$. Now, for every node $N$ of $T$, we replace every formula $\alpha \in X_N$ by $\alpha(p_{v_0}/v_0, \ldots, p_{v_n}/v_n)$. The resulting tree $T^*$ is constructed in accordance with the rules (t1) and (t3) described in Section 2.3. Moreover, the label of every leaf $N$ of $T^*$ contains a subset $Y_N$ which can be transformed, using the transformation rules of $Rs$, into a set $Y_N^*$ of verifiers such that for every $d \in \mathcal{D}, Y_N^* \not\subseteq d$. This means that $Y_N^* \in \mathcal{F}$ (if $Y_N^*$ were not in $\mathcal{F}$, then for some $d \in \mathcal{D}$ we would have $Y_N^* \subseteq d$), and, hence, $T^*$ can be expanded to a refutation tree of $X$. ∎

THEOREM 3.5: *If $Rs$ is a (strong) resolution counterpart of a resolution logic, then it is also a (strong) resolution counterpart of $\langle \mathcal{L}, C_{Rs}\rangle$.*

*Proof:* Suppose that $Rs$ is a counterpart of $\langle \mathcal{L}, C\rangle$. By Lemmas 3.3 and 3.4, $Rs$ is refutationally complete with respect to $\langle \mathcal{L}, C_{Rs}\rangle$. To show that the conditions (r2) and (r3) are also satisfied it suffices to prove that

(a) for every finite $X \subseteq L, C(X) = L$ *iff* $C_{Rs}(X) = L$.

Indeed, since $Rs$ is a resolution counterpart of $\langle \mathcal{L}, C\rangle$, (a) implies (r3); since both $\langle \mathcal{L}, C\rangle$ and $\langle \mathcal{L}, C_{Rs}\rangle$ are compact, (a) also implies $\Theta_C^R = \Theta_{C_{Rs}}^R$. To show (a), let $X \subseteq L$ be finite. Clearly, (a) holds if $X$ is clean in $Rs$ (indeed, $Rs$ satisfies (r1) with respect to both $C$ and $C_{Rs}$). If $X$ is not clean, then let us select a variable renaming substitution $e$ such that $e(X)$ is clean. Then, by Proposition 1.4,

$$C_{Rs}(X) = L \text{ iff } C_{Rs}(e(X)) = L \text{ iff } C(e(X)) = L \text{ iff } C(X) = L.$$

This shows (a) and concludes the proof that $Rs$ is a counterpart of $\langle \mathcal{L}, C_{Rs}\rangle$.

Now, let us assume that $Rs$ is a strong resolution counterpart of $\langle \mathcal{L}, C\rangle$. The satisfaction of (r1) and (r3) has been demonstrated above (note that the proofs of Lemmas 3.3 and 3.4 do not refer to (r2)). The proof of (r2*) is based on Exercise 2.6. Let $\Theta$ be the congruence relation of $\mathcal{L}$ defined by: for every $\alpha, \beta \in L$,

$$\alpha\Theta\beta \text{ iff for every } h \in Hom(\mathcal{L}, M_{Rs}), h(\alpha) = h(\beta).$$

It is an easy exercise to verify that $\Theta \subseteq \Theta_{C_{Rs}}$. So, by Exercise 2.6, $\Theta_C \subseteq \Theta_{C_{Rs}}$, from which (r2*) follows immediately. ∎

There is an important corollary to Theorem 3.5 which states that a resolution proof system is a resolution counterpart of a propositional logic if and only if it is a resolution counterpart of a certain strongly finite logic. This stresses the importance of strongly finite logics in the methodological studies of resolution proof systems.

THEOREM 3.6: *If $Rs$ is a resolution counterpart of $\langle \mathcal{L}, C\rangle$, then:*

(i) $\Theta_C^R = \Theta_{C_{Rs}}^R$;

(ii) $\Theta_C^R$ *is recursive.*

*Proof:* (i) has been demonstrated in the proof of Theorem 3.5. To show (ii), let $M_{Rs} = \langle \mathcal{V}, \mathcal{D}\rangle$ be the matrix induced by $Rs$ and let $n$ be the greater of: the number of verifiers of $Rs$ and the cardinality of $Var(Ver)$. First, we show the following fact.

(a) Suppose $\alpha, \beta \in L$ are not $\Theta_C^R$-congruent and such that $Var(\alpha, \beta) \subseteq \{p_{n+1}, \ldots, p_{n+k}\}$. Then there is a finite set $Y \cup \{\delta(p_n)\} \subseteq L^{(n+k)}$, such that one of $Y \cup \{\delta(p_n/\alpha)\}$ or $Y \cup \{\delta(p_n/\beta)\}$ (but not both) is $C_{Rs}$-consistent.

Let $\alpha, \beta \in L$ be as stated. By (i), there are: $X \cup \{\gamma(p_s)\} \subseteq L$, $d \in \mathcal{D}$, and a valuation $h$ of $\mathcal{L}$ into $M_{Rs}$, such that, say, $h(X \cup \{\gamma(p_s/\beta)\}) \subseteq d$, while $h(\gamma(p_s/\alpha)) \notin d$. We may assume that $s > n + k$. Let $v_0, \ldots, v_m$ be an enumeration of $Ver$ and let $e_h$ be the substitution defined in the following way: $e_h(p_i) = p_i$, for all $p_i \in Var(\alpha, \beta)$, $e_h(p_s) = p_n$, and, for the remaining variables, $e_h(p_i) = p_j$, if $h(p_i) = v_j$. Finally, let $h^*$ the valuation such that: for every $i \leq m, h^*(p_i) = v_i$, $h^*(p_n) = h(p_s)$, and $h^*(p_i) = h(p_i)$, for the remaining variables. Then, for every variable $p_i, h^*(e_h(p_i)) = h(p_i)$. Now, if we let $Y = e_h(X)$ and $\delta(p_n) = e_h(\gamma(p_s))$, then $h^*(Y \cup \{\delta(p_n/\beta)\}) \subseteq d$. By Proposition 1.4, we also have $C_{Rs}(Y, \delta(p_n/\alpha)) = L$, and the proof of (a) has been completed.

Next, we claim that

(b) for every $\alpha, \beta \in L$ we can effectively construct a finite set $W \subseteq L$ and formulas $\alpha^*$ and $\beta^*$ such that $\alpha\ \Theta^R_C\ \beta$ *iff* for every finite set $X \cup \{\gamma(p_n)\} \subseteq W$,

$$C_{Rs}(X, \gamma(p_n/\alpha^*)) = L \Leftrightarrow C_{Rs}(X, \gamma(p_n/\beta^*)) = L.$$

To show (b), let $\alpha$ and $\beta$ be arbitrary formulas, let $k$ be the cardinality of $Var(\alpha, \beta)$, and let $e^*$ be a variable renaming substitution such that $Var(e^*(\alpha), e^*(\beta)) \subseteq \{p_{n+1}, \ldots, p_{n+k}\}$. By Proposition 1.5(iii),

$$\alpha\ \Theta^R_C\ \beta \text{ iff } e^*(\alpha)\ \Theta^R_C\ e^*(\beta).$$

Let $\Theta$ be the congruence of $\mathcal{L}$ defined by: $\gamma_0 \Theta \gamma_1$ if and only if for every $h \in Hom(\mathcal{L}, M_{Rs}), h(\gamma_0) = h(\gamma_1)$. Since $M_{Rs}$ is finite, $\Theta$ is recursive and $L^{(n+k)}/\Theta$ is finite. Hence, we can effectively select a finite set $W \subseteq L^{(n+k)}$ such that for every formula $\kappa \in L^{(n+k)}$ there exists $\gamma \in W$ such that $\kappa \Theta \gamma$. We claim that $W$ and the formulas $\alpha^* = e^*(\alpha)$ and $\beta^* = e^*(\beta)$ satisfy (b).

The only nontrivial part of the proof of (b) is the *if* part. Let us suppose that $\alpha^*$ and $\beta^*$ are not $\Theta^R_C$-congruent. By (a), there exists a finite set $Y \cup \{\delta(p_n)\} \subseteq L^{(n+k)}$ such that, say, $C_{Rs}(Y, \delta(p_n/\alpha^*)) = L$ while $C_{Rs}(Y, \delta(p_n/\beta^*)) \neq L$. Let $X \subseteq W$ be obtained from $Y$ by replacing every formula of $Y$ by a $\Theta$-congruent formula from $W$. Since $\Theta \subseteq \Theta^R_C$, we have $C_{Rs}(X, \delta(p_n/\alpha^*)) = L$ and $C_{Rs}(X, \delta(p_n/\beta^*)) \neq L$. Similarly, let $\gamma(p_n) \in W$ be such that $\gamma(p_n)\Theta\delta(p_n)$. Then, by the definition of $\Theta$, $\gamma(p_n/\alpha^*)\Theta\delta(p_n/\alpha^*)$ and $\gamma(p_n/\beta^*)\Theta\delta(p_n/\beta^*)$. This means that $C_{Rs}(X, \gamma(p_n/\alpha^*)) = L$ and $C_{Rs}(X, \gamma(p_n/\beta^*)) \neq L$. This concludes the proof of (b) from which (ii) follows immediately. ∎

Theorems 3.5 and 3.6 are important technical results. They allow us to conclude that in cases when only refutational properties of logical calculi are relevant, the study of resolution logics can be restricted to the study of logics induced by proper resolution proof systems.

## 3.2 CHARACTERIZATION OF RESOLUTION LOGICS

In this section we prove two theorems of a theoretical nature. The first of these theorems, Theorem 3.7, states certain criteria for a logical calculus to be a resolution logic. The second theorem, Theorem 3.8, provides some necessary and sufficient conditions for a resolution logic to have a strong resolution counterpart. We shall use both of these criteria to show that there are resolution logics which have no strong resolution counterparts.

THEOREM 3.7: *Let $\mathcal{P} = \langle \mathcal{L}, C \rangle$ be a propositional logic. The following conditions are equivalent:*

(i) *$\mathcal{P}$ is a resolution logic;*

(ii) *there exists a finite proper matrix $M$ such that $\langle \mathcal{L}, C_M \rangle$ and $\mathcal{P}$ have the same inconsistent sets;*

(iii) *there is an integer $k \geq 0$ such that:*

  (a) *the quotient algebra $\mathcal{L}^{(k)}/\Theta_C^R$ is finite;*

  (b) *for every finite set $X \subseteq L, C(X) = L$ iff for every substitution $e$ which maps $L$ into $L^{(k)}, C(e(X)) = L$.*

*Proof:* (i) $\Rightarrow$ (ii): Follows immediately from Theorem 3.5 and Lemma 3.1.

(ii) $\Rightarrow$ (iii): Let $M = \langle \mathcal{A}, \mathcal{D} \rangle$ be a finite matrix such that the logics $\mathcal{P}' = \langle \mathcal{L}, C_M \rangle$ and $\mathcal{P}$ have the same inconsistent sets, and let $k$ be any integer that satisfies Proposition 1.10(i) and (ii) with respect to $\mathcal{P}'$. By Proposition 1.5(i), $\Theta_{C_M} \subseteq \Theta_{C_M}^R$ while by (ii), $\Theta_{C_M}^R = \Theta_C^R$. So, by Proposition 1.10(ii), the algebra $\mathcal{L}^{(k)}/\Theta_C^R$ is finite. This shows (iii)(a).

Condition (iii)(b) follows from Propositions 1.4 and 1.10(i).

(iii) $\Rightarrow$ (i): Let $Rs = \langle \mathcal{V}, \mathcal{F} \rangle$ be a resolution proof system on $\mathcal{L}$ constructed in the following way. Suppose that $L^{(k)}/\Theta_C^R = \{s_0, \ldots, s_t\}$. For every $i \leq t$, let $\beta_i \in s_i \cap L^{(k)}$ and let $Ver_{Rs} = \{\beta_0, \ldots, \beta_t\}$ be the base set of $\mathcal{V}$. For every $n$-ary connective $f$, the corresponding operation $\underline{f}$ of $\mathcal{V}$ is defined by: $\underline{f}(\beta_{i_1}, \ldots, \beta_{i_n}) = \beta_i$ *iff* $f(\beta_{i_1}, \ldots, \beta_{i_n})\Theta_C^R\beta_i$. Finally, the family $\mathcal{F}$ consists of all $C$-inconsistent subsets of $Ver_{Rs}$.

From the definition of $Rs$ we immediately conclude that $Rs$ satisfies (r2) and (r3) with respect to $\mathcal{P}$. Let $M = \langle \mathcal{V}, \mathcal{D} \rangle$ be the matrix induced by $Rs$. We begin the proof of (r1) by showing that:

(x) $Rs$ is proper;

(y) the logic $\langle \mathcal{L}, C_M \rangle$ has the same finite inconsistent sets as $\mathcal{P}$.

To show (x), let $X_f \subseteq L$ be finite and $C$-inconsistent, and suppose that for some $d \in \mathcal{D}$ and some $h \in Hom(\mathcal{L}, M), h(L) \subseteq d$. So, $h(X_f) \subseteq d$ and, since $d$ is $C$-consistent, $C(h(X_f)) \neq L$. Let $e_h$ be the substitution such that for every variable $p, e_h(p) = h(p)$. Then, for every $\alpha \in L, e_h(\alpha)\Theta_C^R h(\alpha)$. So, $C(e_h(X_f)) \neq L$ and, by (iii)(b), $C(X_f) \neq L$, which contradicts our choice of $X_f$.

To show (y), let $X \subseteq L$ be finite and $C_M$-consistent. So, for some valuation $h$ of $\mathcal{L}$ into $M$ and for some $d \in \mathcal{D}, h(X) \subseteq d$. Hence, $C(h(X)) \neq L$. Let $e_h$ be the substitution defined as in the proof of (x). Then $C(e_h(X)) \neq L$ and, by Proposition 1.4, $C(X) \neq L$.

Now, suppose that $C(X) \neq L$. Then, by (iii)(b), there is a substitution $e$ mapping $L$ into $L^{(k)}$ and such that $C(e(X)) \neq L$. Let us select a valuation $h_e$ of $\mathcal{L}$ into $M$ so that: for every variable $p, h_e(p)$ and $e(p)$ are $\Theta_C^R$-congruent. By induction on the complexity of a formula we can show that for every $\gamma \in L$, $h_e(\gamma)$ and $e(\gamma)$ are $\Theta_C^R$-congruent. Since $e(X)$ is $C$-consistent, so is $h_e(X)$. Hence, for some maximal consistent set $d$ in $Rs$, $h_e(X) \subseteq d$. By (x), this means that $C_M(X) \neq L$. This finishes the proof of (y). Now, (r1) follows from (x), (y), and Lemmas 3.3 and 3.4. ∎

Theorem 3.7 describes the range of applicability of the theory of resolution proof systems which we develop in this book. By Theorem 3.7(ii), the class of resolution logics contains all finitely-valued logics (i.e., logics defined by single finite and proper logical matrices). Hence, classical propositional logic is a resolution logic and so are all finitely-valued logics of Łukasiewicz and Post. However, the class of resolution logics contains much more than just finitely-valued logics. The intuitionistic propositional logic $\mathcal{P}_I$ is not finitely-valued. However, since $\mathcal{P}_I$ has the same inconsistent sets of formulas as classical propositional logic, by Theorem 3.7(ii), $\mathcal{P}_I$ is a resolution logic. It is an easy exercise to verify that every standard logic intermediate between $\mathcal{P}_I$ and classical logic is also in the class of resolution logics.

The equivalence between conditions (i) and (ii) of Theorem 3.7 tells us that from the refutational point of view, resolution logics can be considered as finitely-valued. In other words, the family of all inconsistent sets of formulas of a resolution logic can be defined using a finite logical matrix. In general, however, no finite collection of such matrices can define the consequence operation of such a logic.

The class of resolution logics with strong resolution counterparts is described in the following theorem.

THEOREM 3.8 (O'Hearn, Stachniak): *A logic $\langle \mathcal{L}, C \rangle$ has a strong resolution counterpart iff there is a natural number k and a subalgebra $\mathcal{A}$ of $\mathcal{L}^{(k)}$ such that:*

(i) *the quotient algebra $\mathcal{A}/\Theta_C$ is finite;*

(ii) *for every finite $X \subseteq L, C(X) = L$ iff for every substitution e which maps L into $|\mathcal{A}|, C(e(X)) = L$.*

*Proof:* Let $\mathcal{P} = \langle \mathcal{L}, C \rangle$ be a propositional logic. First, suppose that $\mathcal{P}$ is a resolution logic and let $Rs = \langle \mathcal{V}, \mathcal{F} \rangle$ be a strong resolution counterpart of $\mathcal{P}$. Let $k$ be the cardinality of $Var(Ver)$ and let

$$A = \{a \in L^{(k)} : a\Theta_C v, \text{ for some } v \in Ver\}.$$

We claim that $A$ is closed under the application of the logical connectives of $\mathcal{L}$. Indeed, let $f$ be an $m$-ary connective, let $a_1, \ldots, a_m \in A$, and let $v_1, \ldots, v_m \in Ver$ be such that for every $1 \leq j \leq m, v_j\Theta_C a_j$. By (r2*), there exists a verifier $v$ such that $f(v_1, \ldots, v_m)\Theta_C v$. Since $\Theta_C$ is a congruence, we also have $f(a_1, \ldots, a_m)\Theta_C f(v_1, \ldots, v_m)$. So, $f(a_1, \ldots, a_m)\Theta_C v$ and the claim has been justified.

Now, (i) follows from the fact that $A$ together with the connectives of $\mathcal{L}$ form a subalgebra of $\mathcal{L}^{(k)}$, and that $A/\Theta_C$ and $Ver$ are of the same cardinality.

To prove (ii), let $X \subseteq L$ be finite. If $C(X) = L$, then, by Proposition 1.4, $C(e(X)) = L$, for every substitution $e$. If $X$ is $C$-consistent, then $C_{Rs}(X) \neq L$ (cf. (a) in the proof of Theorem 3.5). This means that there exist $h \in Hom(\mathcal{L}, M_{Rs}), d \in \mathcal{D}$, and $\alpha \in L$ such that

(a) $h(X) \subseteq d$ and $h(\alpha) \notin d$,

where $M_{Rs}$ is the matrix induced by $Rs$. Let $e_h$ be the substitution such that for every variable $p, e_h(p) = h(p)$. Clearly, $e_h$ maps $L$ into $A$. Moreover, for every $\beta \in L, e_h(\beta)\Theta_C h(\beta)$. By (a) and (r3), $C(h(X)) \neq L$ and, since $\Theta_C \subseteq \Theta_C^R$, $C(e_h(X)) \neq L$. This completes our proof of the first half of this theorem.

For the other half, let us assume that for some integer $k$ and for some subalgebra $\mathcal{A}$ of $\mathcal{L}^{(k)}$, the conditions (i) and (ii) are satisfied. Let $Rs = \langle \mathcal{V}, \mathcal{F} \rangle$ be a resolution proof system on $\mathcal{L}$ defined as in the part (iii)⇒(i) of the proof of Theorem 3.7, with $\mathcal{L}^{(k)}$ replaced by $\mathcal{A}$ and $\Theta_C^R$ replaced by $\Theta_C$. Clearly, $Rs$ satisfies (r2*) and (r3). As in the part (iii)⇒ (i) of the proof of Theorem 3.7 we show that $Rs$ is proper and satisfies (r1).∎

By comparing Theorem 3.7 with Theorem 3.8 we conclude that if a resolution logic $\langle \mathcal{L}, C \rangle$ has no strong resolution counterparts, then for some $k \geq 0$, $L^{(k)}/\Theta_C$ is infinite. This gives us an idea how to construct a resolution logic which does not have strong resolution counterparts and, hence, how to prove that the class of resolution logics is essentially larger than the class of logical systems with strong resolution counterparts.

EXAMPLE 3.9: Let $\mathcal{L}$ be a propositional language with two unary connectives $F$ and $G$. Let $C$ be the consequence operation on $\mathcal{L}$ defined in the following way. For every $X \subseteq L$,

$$C(X) = \begin{cases} L, & \text{if for some } \alpha \in L, F(\alpha) \in X, \\ X, & \text{otherwise.} \end{cases}$$

We claim that $\langle \mathcal{L}, C \rangle$ is a resolution logic without strong resolution counterparts.

First, let us note that $C$ is a structural consequence operation. Moreover, $L^{(0)}/\Theta_C^R$ consists of two elements: the $\Theta_C^R$-equivalence class of $p_0$ and the $\Theta_C^R$-equivalence class of $Fp_0$. Hence, Theorem 3.7(iii)(a) is satisfied for $k = 0$. It is an easy task to verify that Theorem 3.7(iii)(b) is satisfied as well. Indeed, by Proposition 1.4, if $X \subseteq L$ is $C$-inconsistent, then so is $e(X)$, for every endomorphism $e$. Conversely, if $C(X) \neq L$, then $C(e(X)) \neq L$, where $e$ is selected so that $e(p) = p_0$, for every variable $p \in Var(X)$. To conclude, $\mathcal{P}$ is a resolution logic.

Next, let $k \geq 0$ and let $\mathcal{A}$ be a subalgebra of $\mathcal{L}^{(k)}$. If $\gamma$ is an element of $\mathcal{A}$, then the $\Theta_C$-equivalence classes of $F\gamma, FF\gamma, FFF\gamma, \ldots$ are distinct and, hence, $\mathcal{A}/\Theta_C$ is infinite. This shows that the resolution logic $\mathcal{P}$ does not have strong resolution counterparts. ∎

The logic $\mathcal{P}$ constructed in Example 3.9 is a resolution logic which is not strongly finite. Indeed, as we have just shown, for every $k \geq 0, \mathcal{L}^{(k)}/\Theta_C$ is infinite. This, in view of Proposition 1.10, shows that $\mathcal{P}$ cannot be definite by a single finite matrix.

## 3.3 DISJUNCTIVE RESOLUTION LOGICS

Let $Rs$ be a resolution counterpart of a disjunctive logic $\mathcal{P}$ and let $\vee$ be a disjunction connective of $\mathcal{P}$. We assume that $v_0, \ldots, v_n$ is a fixed enumeration of $Ver_{Rs}$. During the discussion on the resolution principle and its interpretation as the rule of verification (Section 2.1), we indicated that the existence of a disjunction connective should allow us to replace the resolution rule by its single-conclusion variant

$$R^{\vee}_{Ver} : \frac{\alpha_0(p), \ldots, \alpha_n(p)}{\alpha_0(p/v_0) \vee \ldots \vee \alpha_n(p/v_n)}.$$

As a result, we should be able to represent refutation proofs in $Rs$ as finite sequences of formulas rather than proof trees. We call $\alpha_0(p/v_0) \vee \ldots \vee \alpha_n(p/v_n)$ the $R^{\vee}_{Ver}$-*resolvent* of $\alpha_0(p/v_0), \ldots, \alpha_n(p/v_n)$ upon $p$.

The main result presented in this section asserts that $Rs$ remains a resolution counterpart of $\mathcal{P}$ (to be exact, (r1)–(r3) remain true), when the rule $R_{Ver}$ is replaced by $R^{\vee}_{Ver}$. Let $M_{Rs} = \langle \mathcal{V}, \mathcal{D} \rangle$ be the matrix induced by $Rs$. If $X$ is a finite set of formulas, then we shall write

$$X \overset{Rs}{\Rightarrow}_{\vee} \Box$$

if there is a finite sequence $\alpha_0, \ldots, \alpha_m$ of formulas (a *refutation of* $X$ in $Rs$) such that:

- each $\alpha_i$ either is in $X$ or is obtained from some formulas in the sequence earlier than $\alpha_i$ with the help of the resolution rule $R^{\vee}_{Ver}$ or a transformation rule of $Rs$;
- some subset $Y$ of $\{\alpha_0, \ldots, \alpha_m\}$ is a member of $\mathcal{F}$.

LEMMA 3.10: *Let $Rs$ be a resolution counterpart of a disjunctive logic and let $v_0 \vee v_1 \Rightarrow v$ be a transformation rule of $Rs$, where $v_0, v_1 \in Ver$. Then, for every $d \subseteq Ver$ which is maximal consistent in $Rs$, $v \in d$ iff $v_0 \in d$ or $v_1 \in d$.*

*Proof:* Let $Rs$ be a resolution counterpart of a disjunctive logic $\langle \mathcal{L}, C \rangle$ and let

$v, v_0, v_1$, and $d$ be as stated. If $v \in d$, then $C(d,v) \neq L$ and, since $v \ \Theta_C^R \ v_0 \vee v_1$, we also have $C(d, v_0 \vee v_1) \neq L$. By disjunctivity of $C$, either $C(d, v_0) \neq L$ or $C(d, v_1) \neq L$. Since $d$ is maximal consistent, we conclude that either $v_0$ or $v_1$ belongs to $d$.

Conversely, let us assume that $v \notin d$. Then $d \cup \{v\}$ is $C$-inconsistent and, hence, $C(d, v_0 \vee v_1) = L$. Since $C$ is disjunctive, we must have $C(d, v_0) = C(d, v_1) = L$ and, since $d$ is $C$-consistent, neither $v_0$ nor $v_1$ is in $d$. ∎

THEOREM 3.11: *If $Rs$ is a resolution counterpart of a disjunctive logic, then the logic defined by the matrix induced by $Rs$ is also disjunctive.*

*Proof:* Let $Rs$ be a resolution counterpart of a disjunctive logic $\langle \mathcal{L}, C\rangle$ and let $C_{Rs}$ denote the consequence operation defined by the matrix $M_{Rs} = \langle \mathcal{V}, \mathcal{D}\rangle$ induced by $Rs$. Let $X \cup \{\alpha, \beta\} \subseteq L$. We want to show that

$$C_{Rs}(X, \alpha \vee \beta) = C_{Rs}(X, \alpha) \cap C_{Rs}(X, \beta).$$

First, let us assume that $\gamma \in C_{Rs}(X, \alpha \vee \beta)$. Let $h \in Hom(\mathcal{L}, M_{Rs})$ and $d \in \mathcal{D}$, be such that $h(X \cup \{\alpha\}) \subseteq d$. By Lemma 3.10, $h(\alpha \vee \beta) \in d$. Since $\gamma \in C_{Rs}(X, \alpha \vee \beta), h(\gamma) \in d$. This shows that $\gamma \in C_{Rs}(X, \alpha)$. In the same way we prove that $\gamma \in C_{Rs}(X, \beta)$.

Conversely, if $\gamma \notin C_{Rs}(X, \alpha \vee \beta)$, then for some $h \in Hom(\mathcal{L}, M_{Rs})$ and some $d \in \mathcal{D}, h(X \cup \{\alpha \vee \beta\}) \subseteq d$ while $h(\gamma) \notin d$. Again, by Lemma 3.10, either $h(\alpha) \in d$ or $h(\beta) \in d$, showing that $\gamma \notin C_{Rs}(X, \alpha) \cap C_{Rs}(X, \beta)$. ∎

From Theorem 3.11 we immediately conclude that the logic $\langle \mathcal{L}, C_{Rs}\rangle$ satisfies the associativity law for disjunction, i.e., for every $\alpha, \beta, \gamma \in L, C_{Rs}(\alpha \vee (\beta \vee \gamma)) = C_{Rs}((\alpha \vee \beta) \vee \gamma)$. This justifies our choice of parenthesis-free notation for $R_{Ver}^{\vee}$-resolvents.

LEMMA 3.12: *Let $Rs$ be a resolution counterpart of a disjunctive logic and let $X \subseteq L$ be finite and clean in $Rs$. If $X \overset{Rs}{\Rightarrow}_{\vee} \Box$, then $C_{Rs}(X) = L$.*

*Proof:* Let

$$\alpha_0, \ldots, \alpha_k, \beta_0, \ldots, \beta_l$$

be a refutation of $X$ in $Rs$ based on $R_{Ver}^{\vee}$. We want to show that $C_{Rs}(X) = L$. To make the proof less complex, let us assume that $\alpha_k$ and $\beta_l$ are verifiers and that $\{\alpha_k, \beta_l\} \in \mathcal{F}$. Moreover, let us assume that $\alpha_0, \ldots, \alpha_k$ is the derivation of $\alpha_k$ and that $\beta_0, \ldots, \beta_l$ is the derivation of $\beta_l$ from $X$.

First, let us concentrate on the sequence $\alpha_0, \ldots, \alpha_k$. Let us assume that no application of a transformation rule precedes an application of $R^{\vee}_{Ver}$. Hence, for some $j < k, \alpha_k$ is obtained from $\alpha_j$ using transformation rules only. Furthermore, we may require that $\alpha_j = \gamma_0 \vee \ldots \vee \gamma_s$ and that every $\gamma_i$, $i \leq s$, is obtained from some $\gamma_i^* \in X$ by substituting verifiers for all the variables of $\gamma_i^*$. (In other words, $\alpha_j$ is obtained from $\gamma_0^*, \ldots, \gamma_s^*$ using only $R^{\vee}_{Ver}$.) Since $Var(\alpha_j) \cap Var(X) = \emptyset$, for every $h \in Hom(\mathcal{L}, M_{Rs})$, there is $m \leq s$ such that $\gamma_m$ is obtained from $\gamma_m^*$ by replacing every variable $p$ occurring in $\gamma_m^*$ with $h(p)$ (see also Lemma 3.17). Let us rename $\gamma_m^*$ as $\alpha_h$. We claim that

(a) for every $d \in \mathcal{D}$, if $\alpha_k \notin d$, then $h(\alpha_h) \notin d$.

To justify (a), let us note that if $\alpha_k \notin d$, then, by Lemma 3.10, every $\gamma_i$ transforms into a verifier not in $d$. Since $h(\alpha_h)\Theta^R_C\gamma_m$ and since $Rs$ does not have $\Theta^R_C$-congruent verifiers, $\gamma_m$ transforms into $h(\alpha_h)$ and, hence, $h(\alpha_h) \notin d$. This shows (a).

In the same way we can prove that for every valuation $h$ there exists a formula $\beta_h \in X$ such that

(b) for every $d \in \mathcal{D}$, if $\beta_l \notin d$, then $h(\beta_h) \notin d$.

Finally, since for no $d \in \mathcal{D}, \{\alpha_k, \beta_l\} \subseteq d$, by (a) and (b) we conclude that for every valuation $h$ and every $d \in \mathcal{D}$ there exist $\alpha_h, \beta_h \in X$ such that $h(\alpha_h) \notin d$ or $h(\beta_h) \notin d$. Hence, $C_{Rs}(X) = L$. ∎

LEMMA 3.13: *Let $Rs$ and $X$ be as in Lemma 3.12. Then $C_{Rs}(X) = L$ implies $X \overset{Rs}{\Rightarrow}_{\vee}\Box$.*

*Proof:* Let $X \subseteq L$ be finite and clean in $Rs$. Our proof uses the semantic tree argument presented in the proof of Lemma 3.4.

Let $T$ be the tree constructed as in the proof of Lemma 3.4 with only one modification. Since $R^{\vee}_{Ver}$ is a single-conclusion rule, all internal nodes of $T$ have just one child. So, if for some $k \geq 1$, $S_0^N, \ldots, S_m^N$ are the closed semantic trees associated with the (only) node $N$ at level $k$, and if for some $j \leq m$, there exists an inference node $K$ of $S_j^N$ whose every child $K_i$ falsifies some formula $\alpha_i(p_{k+1}) \in X_N$, then, by Lemma 3.10, the partial valuation associated with $K$ falsifies the formula $\beta = \alpha_0(p_{k+1}/p_{v_0}) \vee \ldots \vee \alpha_n(p_{k+1}/p_{v_n})$. $\beta$, of course, 'simulates' the resolvent of $\alpha_0(p_{k+1}), \ldots, \alpha_n(p_{k+1})$. So, only one node is attached to $N$ and labelled with

$X_N \cup \{\beta\}$. Other details concerning the construction of $T$ are as in the proof of Lemma 3.4.

Let us assume that $N_0, \ldots, N_s$ are all the nodes of $T$ listed level by level. By the construction of $T$, for every $j < s$, the label of $N_{j+1}$ is obtained from the label of $N_j$ by the addition of some formula $\beta_{j+1}$ (deduced, of course, using an inference rule of $Rs$). So, the formulas of $X$ together with the sequence $\beta_1, \ldots, \beta_s$ form a refutation of $X$ based on $R^{\vee}_{Ver}$. ∎

From Lemmas 3.12 and 3.13 we conclude that:

THEOREM 3.14: *Let Rs be a resolution counterpart of a disjunctive logic and let $X \subseteq L$ be finite and clean in Rs. Then $X \overset{Rs}{\Rightarrow} \square$ iff $X \overset{Rs}{\Rightarrow}_{\vee} \square$.*

By this theorem, the resolution rule of every resolution counterpart of a disjunctive logic can be rewritten as a single-conclusion rule. Let us remark that the same result can be obtained for some non-disjunctive resolution logics (see Exercise 3.8).

In view of Theorem 3.14 and the refutational principle (ref), we can determine the validity of an inference $\alpha \in C(X)$ by using $R^{\vee}_{Ver}$ and the transformation rules of $Rs$ to establish the consistency status of $X_\alpha$. To determine whether or not $X_\alpha$ is consistent, we can, for instance, execute the following *level saturation* algorithm.

(d1) Let $Z$ be the result of simplification of formulas of $X_\alpha$ using the transformation rules of $Rs$. If for some $V \subseteq Z, V \in \mathcal{F}$, then the deductive process terminates successfully and we conclude that $X_\alpha$ is $C$-inconsistent.

(d2) Generate the set $r(X)$ of all $R^{\vee}_{Ver}$-resolvents of formulas of $X_\alpha$. If $r(X) \subseteq X_\alpha$, then the deductive process terminates unsuccessfully and we conclude that $X_\alpha$ is $C$-consistent. Else, we let $X_\alpha$ be $X_\alpha \cup r(X)$ and we return to step (d1).

This algorithm systematically generates and simplifies resolvents until one of the termination conditions are met. Unfortunately, with every execution of step (d2) of this algorithm, the *search space*, i.e., the set $X_\alpha$, becomes larger and larger as new formulas are generated and added to it. Since an uncontrolled growth of $X_\alpha$ can easily result in the combinatoric explosion, an efficient level saturation-like algorithm must employ some strategies which at least restrict the application of $R^{\vee}_{Ver}$ in step (d2). We deal with the issue of the efficiency of the deductive process and, in particular, with theorem proving strategies, in Chapters 4 and 5.

Now, we are ready to present a number of examples of resolution logics and their 'disjunctive' resolution counterparts.

EXAMPLE 3.15: Let us begin with classical propositional logic $\mathcal{P}_2$. In Example 2.2 we have informally described a two-verifier resolution counterpart of this logic. The formal definition of this system, which we denote by $Rs_2$, is as follows. We select the formulas $v_1 = p_0 \to p_0$ and $v_0 = \neg v_1$ as the verifiers. Then,

$$Rs_2 = \langle \mathcal{V}_2, \{\{v_0\}, \{v_0, v_1\}\}\rangle,$$

where $\mathcal{V}_2 = \langle \{v_0, v_1\}, \neg, \vee, \wedge, \to\rangle$, and the operations $\neg, \vee, \wedge$, and $\to$ are defined as follows:

$$\neg v_i \Rightarrow v_{1-i}, \qquad v_i \vee v_j \Rightarrow v_{max(i,j)}, \qquad v_i \wedge v_j \Rightarrow v_{min(i,j)},$$

$$v_i \to v_j \Rightarrow \begin{cases} v_0, & \text{if } i = 1 \text{ and } j = 0, \\ v_1, & \text{otherwise.} \end{cases}$$

The disjunctive resolution rule of $Rs_2$ has the form

$$\frac{\alpha_0(p), \alpha_1(p)}{\alpha_0(p/v_0) \vee \alpha_1(p/v_1)}.$$

It is not difficult to verify that $Rs_2$ is a resolution counterpart of $\mathcal{P}_2$. Conditions (r2) and (r3) can be verified directly and we leave this task to the reader. By Theorem 3.14, to show that $\mathcal{P}_2$ and $Rs_2$ satisfy (r1) it suffices to prove that $\mathcal{P}_2$ and the system $Rs_2^*$ defined in Example 2.3 satisfy (r1). This, however, follows immediately from Lemmas 3.3 and 3.4, and from the fact that the matrix induced by $Rs_2$ defines $\mathcal{P}_2$.

As we shall see shortly, $Rs_2$ is also a resolution counterpart of every logic which has the same inconsistent sets of formulas as $\mathcal{P}_2$ (cf. Theorem 3.19). Hence, $Rs_2$ is a resolution counterpart of the intuitionistic logic $\mathcal{P}_\mathcal{I}$ as well as any standard logic intermediate between $\mathcal{P}_\mathcal{I}$ and $\mathcal{P}_2$. An example of a refutation of a set of formulas in $Rs_2$ can be found in Example 2.1.

Our next two examples of disjunctive resolution logics are the 3- and the 4-valued Łukasiewicz logics. The resolution proof system $Rs_3 = \langle \mathcal{V}_3, \mathcal{F}_3\rangle$ has six verifiers:

$$v_5 = p_0 \to p_0,\ v_4 = p_0 \vee \neg p_0,\ v_3 = \neg(v_4 \to \neg v_4),\ v_2 = \neg v_3,\ v_1 = \neg v_4,\ v_0 = \neg v_5.$$

The transformation rules of $Rs_3$ can be represented as tables. Let us consider the transformation rules for the disjunction connective listed in Table 3.1. The rows, columns, and the values recorded in this table are the indices of verifiers of $Rs_3$. Hence, according to Table 3.1, $v_1 \vee v_0$ transforms into $v_1$ while $v_1 \vee v_4$ into $v_4$. The transformation rules for the remaining connectives can be found in Appendix B, where the complete description of $Rs_3$ is given.

TABLE 3.1: Transformation rules of $Rs_3$ for $\vee$.

| $\vee$ | 0 | 1 | 2 | 3 | 4 | 5 |
|---|---|---|---|---|---|---|
| 0 | 0 | 1 | 2 | 3 | 4 | 5 |
| 1 | 1 | 1 | 2 | 4 | 4 | 5 |
| 2 | 2 | 2 | 2 | 5 | 5 | 5 |
| 3 | 3 | 4 | 5 | 3 | 4 | 5 |
| 4 | 4 | 4 | 5 | 4 | 4 | 5 |
| 5 | 5 | 5 | 5 | 5 | 5 | 5 |

The family $\mathcal{F}_3$ is defined by the following equivalence: for every set $V$ of verifiers of $Rs_3$,

$$V \in \mathcal{F}_3 \quad \textit{iff} \quad C_3(V) = L,$$

and can be determined with the help of $M_3$.

One of the resolution counterparts of the 4-valued Łukasiewicz logic $\mathcal{P}_4$ can be defined as follows. The system $Rs_4 = \langle \mathcal{V}_4, \mathcal{F}_4 \rangle$ has eight verifiers:

$v_7 = p_0 \to p_0$,
$v_6 = p_0 \vee \neg p_0$,
$v_5 = \neg(p_0 \to \neg p_0) \wedge (\neg p_0 \to p_0)$,
$v_4 = \neg(p_0 \to (p_0 \to \neg p_0)) \wedge (\neg p_0 \to (\neg p_0 \to p_0))$,
$v_3 = \neg v_4, \quad v_2 = \neg v_5, \quad v_1 = \neg v_6, \quad v_0 = \neg v_7$.

Table 3.2 lists the transformation rules of $Rs_4$ for $\to$ and $\neg$. The transformation rules for the remaining connectives can be found in Appendix C, where the complete description of $Rs_4$ is given.

TABLE 3.2: Transformation rules of $Rs_4$ for $\rightarrow$ and $\neg$.

| $\rightarrow$ | 0 | 1 | 2 | 3 | 4 | 5 | 6 | 7 |
|---|---|---|---|---|---|---|---|---|
| 0 | 7 | 7 | 7 | 7 | 7 | 7 | 7 | 7 |
| 1 | 6 | 7 | 7 | 7 | 6 | 7 | 7 | 7 |
| 2 | 5 | 6 | 7 | 7 | 5 | 6 | 7 | 7 |
| 3 | 4 | 5 | 6 | 7 | 4 | 5 | 6 | 7 |
| 4 | 3 | 3 | 3 | 3 | 7 | 7 | 7 | 7 |
| 5 | 2 | 3 | 3 | 3 | 6 | 7 | 7 | 7 |
| 6 | 1 | 2 | 3 | 3 | 5 | 6 | 7 | 7 |
| 7 | 0 | 1 | 2 | 3 | 4 | 5 | 6 | 7 |

| | $\neg$ |
|---|---|
| 0 | 7 |
| 1 | 6 |
| 2 | 5 |
| 3 | 4 |
| 4 | 3 |
| 5 | 2 |
| 6 | 1 |
| 7 | 0 |

The family $\mathcal{F}_4$ is defined by the following equivalence: for every set $V$ of verifiers of $Rs_4$,

$$V \in \mathcal{F}_4 \quad \textit{iff} \quad C_4(V) = L.$$

The proofs that $Rs_3$ and $Rs_4$ are counterparts of $\mathcal{P}_3$ and $\mathcal{P}_4$, respectively, are given in Chapter 4. In fact, these resolution proof systems are minimal in the sense that every resolution counterpart of $\mathcal{P}_3$ (of $\mathcal{P}_4$) has at least six (at least eight) verifiers. ∎

## 3.4 RELATIVE SOUNDNESS OF THE RESOLUTION RULE

The refutational completeness principle (r1) is restricted to clean sets only. It turns out that for some logics the unrestricted form of (r1) would eliminate many theoretically important resolution proof systems from the class of resolution counterparts of these calculi. For instance, as the following example shows, the system $Rs_3$ would not be among the counterparts of the 3-valued Łukasiewicz logic.

EXAMPLE 3.16: Let $Rs_3$ and $\mathcal{P}_3$ be as in Example 3.15 and let $X = \{v_2(p_0/p_1), v_4\}$. Clearly, $X$ is consistent in $\mathcal{P}_3$ but it is not clean in $Rs_3$. The application of the resolution rule to the sequence

$$v_2(p_0/p_1),\ v_2(p_0/p_1),\ v_2(p_0/p_1),\ v_2(p_0/p_1),\ v_2(p_0/p_1),\ v_2(p_0/p_1)$$

results in the resolvent $\alpha$ that can be transformed, using the transformation rules of $Rs_3$, into $v_2$. Since $C_3(\{v_2, v_4\}) = L$, $\{v_2, v_4\} \in \mathcal{F}_3$. Hence, if (r1) were not restricted to clean sets only, then $Rs_3$ would fail (r1) and, therefore, would not be a counterpart of $\mathcal{P}_3$. ∎

The above example shows that the resolution rule of some resolution counterparts is unsound, i.e., when applied to some consistent (non-clean) sets of formulas may yield an inconsistent set of resolvents. However, (r1) guarantees that the resolution rule is 'sufficiently' sound, i.e., it is sound when used during the refutation process of clean sets. To prove this fact we need the following lemma.

LEMMA 3.17: *Let Rs be a resolution counterpart of a disjunctive logic. Let $X \subseteq L$ be finite and clean in Rs, let $Var(X) = \{p_1, \ldots, p_k\}$, and let $\rho$ be a mapping from $\{1, \ldots, k\}$ into Ver. If $\alpha$ is deduced from X using only $R^{\vee}_{Ver}$, then there exists $\beta \in X$ such that some disjunct of $\alpha$ is obtained from $\beta$ by replacing some of the variables $p_i$ with the verifiers $\rho(i)$.*

Before moving on to the proof, let us illustrate the main point of this lemma with the following simple example. Let $X$ and $\alpha$ be as stated, let $Var(X) = \{p_1, p_2\}$, and let $Ver = \{v_0, v_1\}$. If the only variables occurring in $\alpha$ are those occurring in verifiers, then, according to Lemma 3.17, there are $\beta_0, \beta_1, \beta_2, \beta_3 \in X$ such that $\alpha$ is

$$\beta_0(p_1/v_0, p_2/v_0) \vee \beta_1(p_1/v_1, p_2/v_0) \vee \beta_2(p_1/v_0, p_2/v_1) \vee \beta_3(p_1/v_1, p_2/v_1).$$

In other words, every possible assignment of verifiers to variables $p_1$ and $p_2$ is represented by a certain disjunct of $\alpha$.

*Proof:* The proof is by induction on the number of applications of the resolution rule used to deduce $\alpha$. Let $X, \rho$, and $\alpha$ be as stated and let

$$\alpha = \alpha_0(p_i/v_0) \vee \ldots \vee \alpha_n(p_i/v_n).$$

Clearly, if for every $j \leq n, \alpha_j(p_i) \in X$, then $\alpha_j(p_i)$ such that $\rho(i) = v_j$ satisfies the conclusion of the lemma.

Now, let us assume that the lemma holds for every $\alpha_j(p_i)$. Consider the disjunct $\alpha_j(p_i/\rho(i))$ of $\alpha$ and let $\beta(p_i) \in X$ be a formula that satisfies the conclusion of this lemma for $\alpha_j(p_i)$. Let $\gamma(p_i)$ be the disjunct of $\alpha_j(p_i)$ obtained from $\beta(p_i)$ by replacing some of the variables $p_j$ with the verifiers $\rho(j)$. Clearly, $\gamma(p_i/\rho(i))$ is a disjunct of $\alpha$ that satisfies the conclusion of the lemma with respect to $\beta(p_i)$. ∎

THEOREM 3.18 Soundness of the Resolution Rule: *Let $Rs$ be a resolution counterpart of a disjunctive logic $\langle \mathcal{L}, C \rangle$ and let $X \subseteq L$ be finite and clean in $Rs$. Let $r(X)$ denote the smallest set including $X$ and closed under the application of the resolution rule $R^{\vee}_{Ver}$. Then $C(X) \neq L$ implies $C(r(X)) \neq L$.*

*Proof:* Let $X \subseteq L$ be finite, clean in $Rs$, and $C$-consistent. Suppose that $Var(X) = \{p_1, \ldots, p_k\}$ and, for reasons of simplicity of presentation, that $Var(Ver) = \{p_0\}$. Let $C_{Rs}$ denote the consequence operation defined by the matrix $M_{Rs}$ induced by $Rs$. By Lemma 3.4, $C_{Rs}(X) \neq L$. So, for some $h \in Hom(\mathcal{L}, M_{Rs})$ and some set $d$ of designated truth-values of $M_{Rs}, h(X) \subseteq d$. Let $\rho$ be the mapping from $\{1, \ldots, k\}$ into $Ver$ satisfying $\rho(i) = h(p_i)$, all $1 \leq i \leq k$. Since for every $\alpha \in X, h(\alpha)\ \Theta^R_C\ \alpha(p_1/\rho(1), \ldots, p_k/\rho(k))$, and since $h(X)$ is $C$-consistent, there is $h^* \in Hom(\mathcal{L}, M_{Rs})$ and a set $d^*$ of designated truth-values of $M_{Rs}$ such that

(a) $h^*(\alpha(p_1/\rho(1), \ldots, p_k/\rho(k))) \in d^*$, all $\alpha \in X$.

Let $\underline{h} \in Hom(\mathcal{L}, M_{Rs})$ be defined as follows: $\underline{h}(p_0) = h^*(p_0)$ and for every $1 \leq j \leq k, \underline{h}(p_j) = h^*(\rho(j))$. We claim that

(b) for every $\alpha \in r(X), \underline{h}(\alpha) \in d^*$.

From (b) we immediately conclude that $C_{Rs}(r(X)) \neq L$ and, since every $C_{Rs}$-consistent set is $C$-consistent (cf. the proof of Theorem 3.5), that $C(r(X)) \neq L$. To show (b), let $\alpha \in r(X)$. If $\alpha \in X$, then, by (a),

$$\underline{h}(\alpha)\ \Theta^R_C\ \alpha(p_1/\underline{h}(p_1), \ldots, p_k/\underline{h}(p_k)) =$$
$$\alpha(p_1/h^*(\rho(1)), \ldots, p_k/h^*(\rho(k)))\Theta^R_C\ h^*(\alpha(p_1/\rho(1), \ldots, p_k/\rho(k))) \in d^*.$$

Since $d^*$ is maximal consistent, $\underline{h}(\alpha) \in d^*$. Finally, if $\alpha = \alpha_0(p_i/v_0) \vee \ldots \vee \alpha_n(p_i/v_n)$, then, by Lemma 3.17, there is $\beta \in X$ such that $\beta^* = \beta(p_1/\rho(1), \ldots, p_i/\rho(i), p_{i+1}, \ldots, p_k)$, obtained from $\beta$ by replacing some of the variables $p_j$ with $\rho(j)$, is a disjunct of $\alpha$. Hence,

$$\underline{h}(\beta^*)\ \Theta^R_C\ \beta(p_1/\underline{h}(\rho(1)), \ldots, p_i/\underline{h}(\rho(i)), p_{i+1}/\underline{h}(p_{i+1}), \ldots, p_k/\underline{h}(p_k)) =$$
$$\beta(p_1/h^*(\rho(1)), \ldots, p_k/h^*(\rho(k)))\ \Theta^R_C\ h^*(\beta(p_1/\rho(1), \ldots, p_k/\rho(k))) \in d^*.$$

Since $d^*$ is maximal consistent, $\underline{h}(\beta^*) \in d^*$ and, by Lemma 3.10, $\underline{h}(\alpha) \in d^*$. ∎

From Example 3.16 and Theorem 3.18 it follows that only finite and clean sets should be subjected to a refutation process. If a finite set $X$ is not clean, then the variables occurring in formulas of $X$ can be uniformly renamed. By Proposition 1.4, this process can be done with a minimal effort and in such a way that the consistency status of $X$ is preserved.

## 3.5 LATTICES OF RESOLUTION LOGICS

Let $\mathcal{L}$ be a propositional language, let $\mathcal{K}$ be the class of all standard logics on $\mathcal{L}$, and let $\Delta$ be the equivalence relation on $\mathcal{K}$ defined as follows: for every $\mathcal{P}_0, \mathcal{P}_1 \in \mathcal{K}$,

$$\mathcal{P}_0 \Delta \mathcal{P}_1 \quad \textit{iff} \quad \mathcal{P}_0 \text{ and } \mathcal{P}_1 \text{ have the same inconsistent sets of formulas.}$$

For every standard logic $\mathcal{P}$, let $\mathcal{K}_{\mathcal{P}}$ denote the equivalence class of $\mathcal{P}$ modulo $\Delta$. From the logical point of view, the calculi of $\mathcal{K}_{\mathcal{P}}$ may be quite different; from the refutational point of view, however, they can be considered indistinguishable. This suggests that the search for an efficient refutational theorem proving algorithm for $\mathcal{P}$ should not be restricted to the study of the logical and computational properties of $\mathcal{P}$. In some cases, other logics in $\mathcal{K}_{\mathcal{P}}$ may prove to be more helpful than $\mathcal{P}$ itself. In this section we test this hypothesis when $\mathcal{P}$ is a resolution logic. We show that the study of the structure of $\mathcal{K}_{\mathcal{P}}$, as well as the study of properties of logical calculi in $\mathcal{K}_{\mathcal{P}}$, may provide us with helpful information during the search for resolution counterparts of $\mathcal{P}$ with desired properties.

If $\mathcal{P}$ is a resolution logic, then we call $\mathcal{K}_{\mathcal{P}}$ the *resolution class* of $\mathcal{P}$. The following fact is an immediate consequence of the definitions of $\Delta$ and of a resolution counterpart of a propositional logic.

THEOREM 3.19: *If $\mathcal{P}$ is a resolution logic, then so are all the logics in $\mathcal{K}_{\mathcal{P}}$. Moreover, all the logics in $\mathcal{K}_{\mathcal{P}}$ have the same resolution counterparts.*

In view of this theorem, we can use the term 'resolution counterpart of $\mathcal{K}_{\mathcal{P}}$' as synonymous to 'resolution counterpart of $\mathcal{P}$'.

Let $\mathcal{L}$ be a propositional language, fixed for the remainder of this section. The resolution class $\mathcal{K}_{\mathcal{P}}$ of a resolution logic $\mathcal{P}$ has many interesting properties. For instance, by Theorem 3.5, every such class contains a strongly finite logic. In addition, the logics in $\mathcal{K}_{\mathcal{P}}$ form a lattice under the ordering $\leq$ which is defined in the following way. If $\mathcal{P}_0 = \langle \mathcal{L}, C_0 \rangle$ and $\mathcal{P}_1 = \langle \mathcal{L}, C_1 \rangle$ are two logics on $\mathcal{L}$, then

$$\mathcal{P}_0 \leq \mathcal{P}_1 \textit{ iff } \text{for every } X \subseteq L, C_0(X) \subseteq C_1(X).$$

We shall frequently write $C_0 \leq C_1$ instead of $\mathcal{P}_0 \leq \mathcal{P}_1$. It is well known that the set of all structural logics on $\mathcal{L}$ forms a complete lattice under $\leq$, while the set of all standard logics on $\mathcal{L}$ forms a sublattice of the lattice of all structural logics (cf. [79]).

THEOREM 3.20: *For every resolution logic* $\mathcal{P}$, $\langle \mathcal{K}_\mathcal{P}, \leq \rangle$ *is a sublattice of the lattice of all standard logics on* $\mathcal{L}$. *Moreover, for every set* $\Phi \subseteq \mathcal{K}_\mathcal{P}$, $sup(\Phi) \in \mathcal{K}_\mathcal{P}$.

By $sup(\Phi)$ and $inf(\Phi)$ we denote the *least upper bound* and *greatest lower bound* of $\Phi$ in the lattice of all structural logics on $\mathcal{L}$.

*Proof:* Let $\mathcal{P} = \langle \mathcal{L}, C \rangle$ be a resolution logic. As remarked earlier, all standard logics on $\mathcal{L}$ form a lattice under $\leq$. If $\mathcal{P}_i = \langle \mathcal{L}, C_i \rangle, i = 0, 1$, are any standard logics, and if $\langle \mathcal{L}, C_{inf} \rangle$ denotes $inf(\{\mathcal{P}_0, \mathcal{P}_1\})$, then for every $X \subseteq L, C_{inf}(X) = C_0(X) \cap C_1(X)$. This means that if $\mathcal{P}_i$ are in $\mathcal{K}_\mathcal{P}$, then so is $\langle \mathcal{L}, C_{inf} \rangle$.

Let $\Phi \subseteq \mathcal{K}_\mathcal{P}$. Since all the logics in $\Phi$ are standard, so is $sup(\Phi)$ (cf. [79], Theorem 1.5.7). To complete the proof of this theorem we must show that $sup(\Phi)$ has the same inconsistent sets as $\mathcal{P}$. Let $\langle \mathcal{L}, C_{sup} \rangle = sup(\Phi)$. Clearly, $C(X) = L$ implies $C_{sup}(X) = L$. So, let us select $X \subseteq L$ such that $C_{sup}(X) = L$. To show that $C(X) = L$ we use the following fact proved in [79] (cf. [79], Theorem 1.5.7(b)). For every $X \cup \{\alpha\} \subseteq L$,

(a) $\alpha \in C_{sup}(X)$ *iff* there is a finite sequence $\langle \mathcal{L}, C_0 \rangle, \langle \mathcal{L}, C_1 \rangle, \ldots, \langle \mathcal{L}, C_n \rangle$ of logics from $\Phi$ such that $\alpha \in C_0(C_1(\ldots C_n(X)\ldots))$.

Let $X_f \subseteq L$ be finite and $C$-inconsistent. Then $X_f \subseteq C_{sup}(X)$ and, by (a), there is a finite sequence $\langle \mathcal{L}, C_0 \rangle, \langle \mathcal{L}, C_1 \rangle, \ldots, \langle \mathcal{L}, C_n \rangle$ of logics from $\Phi$ such that $X_f \subseteq C_0(C_1(\ldots C_n(X)\ldots))$. This means that $C_n(X) = L$ and, hence, we must have $C(X) = L$. ■

The following definition stems from Theorem 3.20. If $\mathcal{P}$ is a resolution logic, then $\langle \mathcal{K}_\mathcal{P}, \leq \rangle$ is called a *resolution lattice*. Every resolution lattice has the least and the greatest element (denoted by $\bot$ and $\top$, respectively). The consequence operation $C_\bot$ of the least element $\bot$ is defined as follows: for every $X \subseteq L$

$$C_\bot(X) = \begin{cases} X, & \text{if } X \text{ is consistent in } \mathcal{P}, \\ L, & \text{otherwise.} \end{cases}$$

The verification that $\bot$ is indeed the least element of $\langle \mathcal{K}_{\mathcal{P}}, \leq \rangle$ does not present any difficulty. The description of $\top$ is given in the following theorem (cf. Figure 3.3).

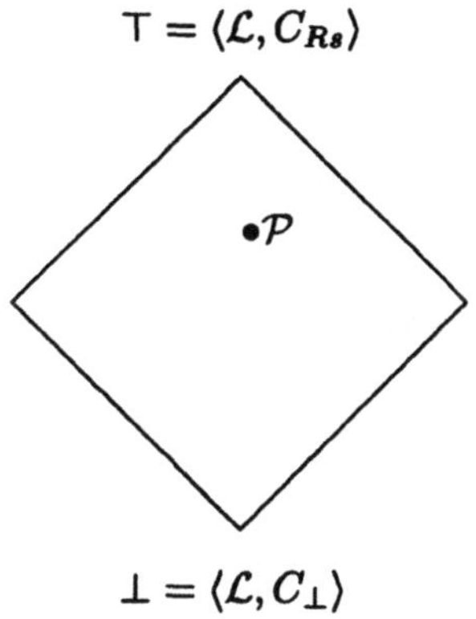


Fig. 3.3: Resolution lattice defined by $\mathcal{P}$.

THEOREM 3.21: *If $Rs$ is a resolution counterpart of a logic $\mathcal{P}$, then the logic $\langle \mathcal{L}, C_{Rs} \rangle$ induced by $Rs$ is the greatest logic in $\mathcal{K}_{\mathcal{P}}$ (with respect to $\leq$).*

*Proof:* Let $\mathcal{P}, Rs$, and $\mathcal{K}_{\mathcal{P}}$ be as stated. Let $\mathcal{P}^* = \langle \mathcal{L}, C^* \rangle$ denote the greatest element of the lattice $\langle \mathcal{K}_{\mathcal{P}}, \leq \rangle$. By Theorem 3.5,

(a) $C_{Rs} \leq C^*$.

To show that also $C^* \leq C_{Rs}$, let us select $X \cup \{\alpha\} \subseteq L$ such that $\alpha \in C^*(X)$, and let us assume that $C_{Rs}(X) \neq L$. This means that there is a valuation $h$ of $\mathcal{L}$ into $M_{Rs}$, and a set $d$ of designated truth-values of $M_{Rs}$, such that $h(X) \subseteq d$. We want to show that $h(\alpha) \in d$. Let $e_h$ be the substitution such that for every variable $p$, $e_h(p) = h(p)$. By Exercise 3.1,

(b) for every $\beta \in L, e_h(\beta)\Theta^R_{C_{Rs}} h(\beta)$.

Since $d$ is maximal consistent and $h(X) \subseteq d$, we have $C_{Rs}(d, e_h(X)) \neq L$. So, $C^*(d, e_h(X)) \neq L$. Since $C^*$ is structural, $e_h(\alpha) \in C^*(d, e_h(X))$. Thus,

(c) $C^*(d, e_h(\alpha)) \neq L$.

Since $\Theta^R_{C_{Rs}} = \Theta^R_{C^*}$, from (b) and (c) we obtain $C^*(d, h(\alpha)) \neq L$. Since $d$ is maximal consistent in $Rs$, we must have $h(\alpha) \in d$, as required. To conclude, $C^* \leq C_{Rs}$ which, together with (a), shows that $C^* = C_{Rs}$. ∎

Theorem 3.21 has a number of simple but useful corollaries which we list below.

COROLLARY 3.22: *If $Rs$ is a resolution counterpart of $\langle \mathcal{L}, C \rangle$, then $C \leq C_{Rs}$.*

COROLLARY 3.23: *If $Rs_0$ and $Rs_1$ are resolution counterparts of a logic $\mathcal{P}$, then $C_{Rs_0} = C_{Rs_1}$.*

The proofs of Corollaries 3.22 and 3.23 follow directly from Theorem 3.21.

If $Rs$ is a resolution counterpart of a logic $\mathcal{P}$, then the logical and computational properties of $Rs$ are determined not solely by $\mathcal{P}$ but by other logics of $\mathcal{K}_{\mathcal{P}}$. For instance, from Theorems 3.11 and 3.14 we conclude that:

THEOREM 3.24: *If $Rs$ is a resolution counterpart of $\mathcal{P}$ and if $\mathcal{K}_{\mathcal{P}}$ has a disjunctive logic, then $Rs$ is a disjunctive resolution counterpart of $\mathcal{P}$, i.e., the resolution rule $R_{Ver}$ of $Rs$ can be replaced with $R^{\vee}_{Ver}$.*

In other words, if a resolution lattice has a disjunctive logic, then every logic in this lattice has a disjunctive resolution counterpart in spite of the fact that some of them may not be disjunctive at all. We shall discuss similar properties of resolution lattices in Chapter 4.

**EXERCISES**

3.1. Let $\mathcal{P} = \langle \mathcal{L}, C \rangle$ be a resolution logic and let $M$ be the matrix induced by a resolution counterpart of $\mathcal{P}$. Let $h \in Hom(\mathcal{L}, M)$ and let $e_h$ be the substitution such that for every variable $p$, $e_h(p) = h(p)$. Show that for every $\alpha \in L$, $e_h(\alpha)\ \Theta^R_C\ h(\alpha)$.

3.2. Let $Rs$ be a proper resolution proof system on $\mathcal{L}$ and let $C_{Rs}$ denote the consequence operation defined by the matrix induced by $Rs$. Show that $\Theta^R_{C_{Rs}}$ is invariant *iff* $\Theta^R_{C_{Rs}} = \Theta_{C_{Rs}}$.

3.3. Let $M = \langle \mathcal{V}, \mathcal{D} \rangle$ be the matrix induced by a proper resolution proof system on $\mathcal{L}$. Show that if $C_M(\emptyset) \neq \emptyset$, then $\bigcap \mathcal{D}$ contains a tautology of $C_M$, i.e., that

$\bigcap \mathcal{D} \cap C_M(\emptyset) \neq \emptyset$.

3.4. Let $\mathcal{L}$ be the language of classical propositional logic. Show that there exists a proper resolution proof system on $\mathcal{L}$ which is a counterpart of no resolution logic.

3.5. Let $\mathcal{L}$ be the language of classical propositional logic and let

$$M_\infty = \langle\langle A_\infty, \underline{\neg}, \underline{\vee}, \underline{\wedge}, \underline{\rightarrow}\rangle, \{1\}\rangle$$

be the matrix for $\mathcal{L}$, where $A_\infty$ is the set of all rational numbers $x$ such that $0 \leq x \leq 1$. The operations $\underline{\neg}, \underline{\vee}, \underline{\wedge}, \underline{\rightarrow}$ are defined as for Łukasiewicz matrices $M_n$, i.e., for every $0 \leq x, y \leq 1$,

$$\underline{\neg} x = 1 - x, \qquad x \underline{\vee} y = max(x, y),$$
$$x \underline{\wedge} y = min(x, y), \qquad x \underline{\rightarrow} y = min(1, 1 - x + y).$$

Show that the logic defined by $M_\infty$ is not a resolution logic.

3.6. Show that there is an effective way to determine whether or not a proper resolution proof system is a resolution counterpart of the logic it induces. Show that the same result holds for strong resolution counterparts.

3.7. Let $Rs$ be a resolution counterpart of a logic $\mathcal{P} = \langle \mathcal{L}, C\rangle$ such that $\Theta_C = \Theta_C^R$. Show that for every standard logic $\mathcal{P}^* = \langle \mathcal{L}, C^*\rangle$ such that $\mathcal{P} \leq \mathcal{P}^* \leq \mathcal{P}_{Rs}$, $\mathcal{P}^*$ is in $\mathcal{K}_\mathcal{P}$ and $\Theta_C = \Theta_{C^*} = \Theta_{C^*}^R$.

3.8. Let $Rs$ be a resolution counterpart of $\mathcal{P} = \langle \mathcal{L}, C\rangle$. Moreover, suppose that $\vee$ is a binary connective such that for every $X \cup \{\alpha, \beta\} \subseteq L$,

$$C(X, \alpha \vee \beta) = L \text{ iff } C(X, \alpha) = L \text{ and } C(X, \beta) = L.$$

(Note that $\vee$ is not necessarily a disjunction.) Show that for every finite $X \subseteq L$ clean in $Rs$,

$$C(X) = L \text{ iff } X \overset{Rs}{\Rrightarrow}_\vee \Box.$$

# 4

# EFFICIENCY OF THE DEDUCTIVE PROCESS

The refutational completeness property (r1) guarantees that the consistency status of every finite and clean set of formulas can be determined after finitely many applications of the inference rules. This property, however, does not provide us with any information as to how the resolution and the transformation rules should be utilized efficiently during the deductive process.

A reasoning program that systematically generates and simplifies resolvents quickly accumulates too much information in its database. Very many of the generated resolvents are already in the database or are irrelevant for the discovery of a computer proof. An uncontrolled growth of the search space can quickly result in combinatoric explosion making the completion of a reasoning task unattainable in an acceptable amount of computer time.

The efficiency of a resolution based reasoning program depends on our ability to control (to minimize) the number of possible applications of the resolution rule during the deductive process. To a certain extent, this can be achieved, among other ways, by minimizing the number of verifiers required by the resolution rule and by using strategies for restricting and directing the application of this rule. In addition, new efficient methods of testing for the termination of the deductive process can be applied.

In this chapter, we discuss ways of minimizing the number of verifiers in the resolution rule. We also propose a new efficient method of testing for the termination of the deductive process and describe a class of resolution logics to which this method is applicable. We shall discuss theorem proving strategies in the next chapter.

## 4.1 MINIMAL RESOLUTION COUNTERPARTS

We say that a resolution proof system $Rs$ on $\mathcal{L}$ is a *minimal resolution counterpart* of a logic $\mathcal{P}$ if no resolution counterpart of $\mathcal{P}$ has fewer verifiers than $Rs$. In this section we discuss the problem of the effective construction of minimal resolution counterparts of resolution logics.

The *size of a resolution proof system* $Rs$, that is the number $card(Ver)$ of its verifiers, is among the main factors that contribute to the complexity of the deductive process. Clearly, the number of possible applications of the resolution rule to formulas of the deduced set grows with the number of verifiers required by this rule. It is also evident that the degree of refutation trees in $Rs$ (or the length of resolvents if $R^{\vee}_{Ver}$ is used) depends on $card(Ver)$ and, hence, so does the number of transformation rules required to complete a reasoning task. Therefore, the search for minimal resolution counterparts of resolution logics is of considerable importance.

First, let us estimate the size of a minimal resolution counterpart of a resolution logics. To this end, let $Rs$ be a resolution counterpart of a logic $\mathcal{P}$ and let $M$ be the smallest proper matrix such that the logic $\langle \mathcal{L}, C_M \rangle$ is in the resolution class $\mathcal{K}_{\mathcal{P}}$ of $\mathcal{P}$. (Let us note that by the criterion of Theorem 3.7, every resolution class contains a strongly finite logic.) The size $card(Ver)$ of $Rs$ can be estimated in terms of the cardinality of $M$ and the number of different variables occurring in the verifiers of $Rs$. (See also Exercise 4.4.)

THEOREM 4.1: *Let $Rs$ be a resolution counterpart of a logic $\mathcal{P} = \langle \mathcal{L}, C \rangle$, and let $t$ denote the number of different variables occurring in the verifiers of $Rs$. Moreover, let $m$ be the smallest integer such that for some proper matrix $M$ of cardinality $m$, $\langle \mathcal{L}, C_M \rangle$ belongs to $\mathcal{K}_{\mathcal{P}}$. Then $m \leq card(Ver) \leq m^{m^t}$.*

*Proof:* Let $\mathcal{P}, Rs, M = \langle \mathcal{A}, \mathcal{D} \rangle$, and $m$ be as stated, and let $p_0, \ldots, p_{t-1}$ be the list of all different variables occurring in the verifiers of $Rs$. By Theorem 3.19, $Rs$ is a resolution counterpart of the logic $\langle \mathcal{L}, C_M \rangle$. For every verifier $v$ of $Rs$, let $f_v : |\mathcal{A}|^t \to |\mathcal{A}|$ be the function defined as follows: for every $h \in Hom(\mathcal{L}, M)$,

$$f_v(h(p_0), \ldots, h(p_{t-1})) = h(v).$$

We claim that

(a) for every two verifiers $v, w$, $v \neq w$ implies $f_v \neq f_w$.

Suppose $v$ and $w$ are two different verifiers of $Rs$. As $Rs$ does not have $\Theta_C^R$-congruent verifiers, for some finite set $X \cup \{\gamma(p)\}$ of formulas we must have, say,

$$C_M(X, \gamma(p/v)) \neq L \text{ and } C_M(X, \gamma(p/w)) = L.$$

This means that for some valuation $h$ of $\mathcal{L}$ into $M$ and for some set $d \in \mathcal{D}$,

$$h(\gamma(p/v)) \in d \text{ and } h(\gamma(p/w)) \notin d.$$

Hence, $h(v) \neq h(w)$ and $f_v(h(p_1), \ldots, h(p_t)) \neq f_w(h(p_1), \ldots, h(p_t))$. This shows (a) from which we immediately conclude that $card(Ver) \leq m^{m^t}$.

By Lemma 3.1, the matrix $M_{Rs}$ induced by $Rs$ is proper. Since the logic defined by $M_{Rs}$ is in $\mathcal{K}_\mathcal{P}$, by the choice of $M$, we must have $m \leq card(M_{Rs}) = card(Ver)$. This completes the proof of the theorem. ∎

COROLLARY 4.2: *The resolution proof system $Rs_2$ is a minimal resolution counterpart of classical propositional logic.*

*Proof:* In Example 3.15 we proved that $Rs_2$ is a resolution counterpart of $\mathcal{P}_2$. Minimality follows from the fact that the smallest matrix satisfying Theorem 4.1 is the two-element matrix defining $\mathcal{P}_2$. ∎

By Theorem 4.1, the size of a minimal resolution counterpart of a resolution logic is between $m$ and $m^{m^t}$ (where $m$ and $t$ are as in Theorem 4.1). As the next theorem shows, $t$ can be arbitrarily large.

THEOREM 4.3: *For every natural number $t > 0$ there exists a resolution logic $\mathcal{P}$ such that the number of different variables occurring in verifiers of any minimal resolution counterpart of $\mathcal{P}$ is $\geq t$.*

PROOF: Let $t > 0$ and let $k = 2t + 3$. Let $\mathcal{L}$ be a propositional language with one unary connective $\neg$, one binary connective $\vee$, and one $k$-ary connective $f$. Finally, let $M$ be the logical matrix for $\mathcal{L}$ with the set $\{1, \ldots, k\}$ of truth-values and with the functions $\underline{\neg}$, $\underline{\vee}$, and $F$ defined as follows: for every $x_1, \ldots, x_k \in \{1, \ldots, k\}$,

$$\underline{\neg}(x_1) = k + 1 - x_1;$$

$$\underline{\vee}(x_1, x_2) = max(x_1, x_2);$$

$$F(x_1,\ldots,x_k) = \begin{cases} 1, & \text{if for some } i \neq j, x_i = x_j, \\ k, & \text{otherwise.} \end{cases}$$

The only designated truth-value of $M$ is $k$. One can easily verify that the logic $\mathcal{P} = \langle \mathcal{L}, C_M \rangle$ is disjunctive and has finite consistent and inconsistent sets of formulas.

Let $Rs$ be a minimal resolution counterpart of $\mathcal{P}$ and let $t_{\mathcal{P}}$ denote the number of different variables occurring in the verifiers of $Rs$. We claim that $t_{\mathcal{P}} \geq t$. For, suppose that $t_{\mathcal{P}} < t$. We need the following lemma.

Lemma A: *Let $n < t$, let $\alpha \in L^{(n)}$, and let $h$ be a valuation of $\mathcal{L}$ into $M$. Then $h(\alpha) \in \{1, k, h(p_0), \ldots, h(p_n), k+1-h(p_0), \ldots, k+1-h(p_n)\}$.*

*Proof:* Simple induction on the complexity of $\alpha$. □

To finish the proof of the theorem, let $C$ denote the consequence operation determined by the matrix $M_{Rs}$ induced by $Rs$. Since $C_M(f(p_1,\ldots,p_k)) \neq L$, by Lemma 3.1(ii), for some $h \in Hom(\mathcal{L}, M_{Rs})$, $C_M(h(f(p_1,\ldots,p_k))) \neq L$. Since $h(f(p_1,\ldots,p_k)) \; \Theta_C^R \; f(h(p_1),\ldots,h(p_k))$ and since $\Theta_C^R = \Theta_{C_M}^R$ (cf. Theorem 3.6), $C_M(f(h(p_1),\ldots,h(p_k))) \neq L$. This means that for some valuation $h'$ of $\mathcal{L}$ into $M$,

$$h'(f(h(p_1),\ldots,h(p_k))) = F(h'(h(p_1)),\ldots,h'(h(p_k))) = k.$$

By the definition of $F$, for every $i \neq j, h'(h(p_i)) \neq h'(h(p_j))$. This, however, contradicts Lemma A. ■

Our next objective is to show that minimal resolution counterparts of resolution logics can be constructed effectively. One way a minimal resolution counterpart $Rs$ of a resolution logic $\mathcal{P} = \langle \mathcal{L}, C \rangle$ can be found is to focus the search for the verifiers of $Rs$ on a sublanguage $\mathcal{L}^{(t)}$ of $\mathcal{L}$, where $t$ is sufficiently large to guarantee that $Ver_{Rs} \subseteq L^{(t)}$. We can use the congruence $\Theta_C^R$ to partition $L^{(t)}$ into finitely many blocks $A_0,\ldots,A_s$ (i.e., congruence classes of $\Theta_C^R$). If we select a formula $\alpha_i$ from every block $A_i$, then, by Theorem 2.11, we may assume that the verifiers of $Rs$ are among the formulas $\alpha_0,\ldots,\alpha_s$. Hence, we can find a minimal resolution counterpart of $\mathcal{P}$ simply by listing all resolution proof systems on $\mathcal{L}$ constructed from $\alpha_0,\ldots,\alpha_s$, and by eliminating those systems from the list which are not counterparts of $\mathcal{P}$. If $t$ is chosen correctly, then the resulting list is finite and contains at least one minimal counterpart of $\mathcal{P}$. (A similar technique can be used to show that minimal strong resolution counterparts of propositional logics can be effectively constructed, and the reader is referred to [66] for more information.) Let us describe this algorithm in more detail.

LEMMA 4.4: *There is an effective method of verifying whether or not a proper resolution proof system is a resolution counterpart of a strongly finite logic.*

*Proof:* Let $Rs$ be a proper resolution proof system on $\mathcal{L}$ and let $M$ be a finite proper matrix for $\mathcal{L}$. By Lemmas 3.3 and 3.4, $Rs$ and $\langle \mathcal{L}, C_M \rangle$ satisfy (r1) if and only if $C_M$ and the consequence operation $C$ defined by $M_{Rs}$ have the same inconsistent sets. Let $N_\times$ denote the direct product $M \times M_{Rs}$ and let $N_\otimes$ denote the ramified product $M \otimes M_{Rs}$. Since $Rs$ and $M$ are proper, by Proposition 1.12,

(a) $C_M$ and $C$ have the same inconsistent sets *iff* the consequence operations defined by $N_\times$ and $N_\otimes$ are equal.

By Proposition 1.11, the truth-value of the right-hand side of (a) can be determined effectively.

If $Rs$ and $\langle \mathcal{L}, C_M \rangle$ satisfy (r1), then $\Theta^R_{C_M} = \Theta^R_C$. By inspecting the proof of Theorem 3.6(ii) we conclude that $\Theta^R_C$ is recursive. Thus, $\Theta^R_{C_M}$ is recursive and (r2) can be verified effectively.

Finally, since $M$ is finite, (r3) can be verified in a finite number of steps. ∎

THEOREM 4.5 Minimization Algorithm: *There is an effective method which for every resolution logic $\mathcal{P}$ and every natural number $t$, constructs a resolution proof system which is minimal among those resolution counterparts $Rs$ of $\mathcal{P}$ for which $Ver_{Rs} \subseteq L^{(t)}$, if such counterparts exist.*

*Proof:* Let $\mathcal{P}$ be a resolution logic and $t$ be a natural number. Let $M$ be a finite proper matrix of cardinality $m$ such that the logic defined by $M$ is in $\mathcal{K}_\mathcal{P}$. Finally, let $\Theta$ be the congruence of $\mathcal{L}^{(t)}$ defined by: for every $\alpha, \beta \in L^{(t)}$,

$$\alpha \Theta \beta \text{ iff for every } h \in Hom(\mathcal{L}, M), h(\alpha) = h(\beta).$$

The finiteness of $M$ implies that $L^{(t)}/\Theta$ is finite and that $\Theta$ is recursive. Hence, we can effectively construct a sequence

(a) $\alpha_0, \ldots, \alpha_s$

of formulas of $L^{(t)}$ such that for every $\beta \in L^{(t)}$ there is $i \leq s$ such that $\beta \Theta \alpha_i$. (In other words, the formulas in (a) are representatives of all congruence classes of $\Theta$.)

We assume that all constant symbols of $\mathcal{L}$ (if such constants exist) are among the formulas of (a). Let us list all proper resolution proof systems on $\mathcal{L}$ whose verifiers are among the formulas listed in (a). Finally, we eliminate from the list in question all these systems which are not counterparts of $\mathcal{P}$. If the resulting list is non-empty, we return a resolution proof system with the smallest number of verifiers.

The correctness of the above algorithm can be justified as follows. Let $Rs$ be a resolution counterpart of $\mathcal{P}$ such that $Var(Ver)$ is of cardinality $\leq t+1$. Since $\Theta \subseteq \Theta^R_{C_M}$, by Corollary 2.13, we can assume that $Ver \subseteq \{\alpha_0, \ldots, \alpha_s\}$. By Lemma 4.4, we can effectively verify whether or not $Rs$ is a counterpart of $\langle \mathcal{L}, C_M \rangle$ (and, hence, a counterpart of $\mathcal{P}$). ∎

In the minimization algorithm of Theorem 4.5, a resolution logic $\mathcal{P} = \langle \mathcal{L}, C \rangle$ is represented by a finite proper matrix $M$ such that $C_M$ and $C$ have the same inconsistent sets. For example, if $\mathcal{P}$ is strongly finite, then $M$ can be any finite proper matrix that defines $\mathcal{P}$. Alternatively, $\mathcal{P}$ can be represented in this minimization algorithm by any of its resolution counterparts (cf. Theorem 3.21), or even by $\Theta^R_C$ and a constant $k$ satisfying Theorem 3.7(iii).

To find a minimal resolution counterpart of a resolution logic $\mathcal{P}$ using the minimization algorithm presented in the proof of Theorem 4.5, we must first solve the problem of finding a constant $t$ which guarantees that the verifiers of some minimal resolution counterpart of $\mathcal{P}$ are in $L^{(t)}$. We solve this problem for resolution logics satisfying $\Theta^R_C = \Theta_C$.

THEOREM 4.6: *There is an algorithm which for every resolution logic $\mathcal{P} = \langle \mathcal{L}, C \rangle$ which satisfies $\Theta_C = \Theta^R_C$, selects a constant $t \geq 0$ such that the verifiers of some minimal resolution counterpart of $\mathcal{P}$ are in $L^{(t)}$.*

*Proof:* Let $\mathcal{P} = \langle \mathcal{L}, C \rangle$ be a resolution logic such that $\Theta_C = \Theta^R_C$, and let $M$ be a finite proper matrix of cardinality $m$ such that the logic defined by $M$ has the same inconsistent sets as $\mathcal{P}$. Let $Rs = \langle \mathcal{V}, \mathcal{F} \rangle$ be a minimal resolution counterpart of $\mathcal{P}$, let $\langle \mathcal{V}, \mathcal{D} \rangle$ be the matrix induced by $Rs$, and let $p_0, \ldots, p_s$ be the list of all variables occurring in verifiers of $Rs$. In view of Exercise 4.4 we can effectively find an integer $k$ such that $Rs$ has at most $k$ verifiers. Our objective is to show that the integer $t = m^{2^k+k-1}$ has the property stated in the theorem. Let us assume that $s > t$. To prove the theorem, we shall construct a substitution $e$ mapping $L$ into $L^{(t)}$ which, when applied to the verifiers of $Rs$, yields a minimal resolution counterpart of $\mathcal{P}$. The construction of $e$ proceeds in two steps.

STEP 1: First, we construct a partition $\Psi$ of the set $\{p_0, \ldots, p_s\}$ into non-intersecting sets, called the members of $\Psi$. If in every verifier of $Rs$ we (uniformly) identify variables belonging to the same element of $\Psi$, then $\mathcal{V}$ will be transformed into the algebra isomorphic to $\mathcal{V}$. The construction of $\Psi$ goes as follows. For every pair $v_i, v_j$ of different verifiers of $Rs$, there exists a valuation $h_{ij}$ of $\mathcal{L}$ into $M$ such that $h_{ij}(v_i) \neq h_{ij}(v_j)$ (if $v_i$ and $v_j$ were not distinguishable in $M$ with the help of a valuation, then they would be $\Theta_C$-congruent). The valuation $h_{ij}$ determines a partition $\Psi_{ij}$ of the set $\{p_0, \ldots, p_s\}$ into at most $m$ disjoint subsets (called the members of $\Psi_{ij}$) so that $p_k$ and $p_r$ belong to the same member of $\Psi_{ij}$ if $h_{ij}(p_k) = h_{ij}(p_r)$. Intuitively, $\Psi_{ij}$ shows how to identify variables of $v_i$ and $v_j$, so that the resulting formulas remain distinguishable in $M$. Clearly, it suffices to identify variables within each member of $\Psi_{ij}$.

Our next objective is to find a minimal set of partitions that separate all the verifiers of $Rs$ in $M$. To this end consider an arbitrary pair $v_i, v_j$ of verifiers and its partition $\Psi_{ij}$. $\Psi_{ij}$ induces a partition of $Ver_{Rs}$ into at least two nonempty sets by grouping verifiers which cannot be separated when the variables in each member of $\Psi_{ij}$ are identified. Next, consider an arbitrary pair $v_r, v_k$ from one of the members $V$ of the partition of $Ver_{Rs}$ and a corresponding partition $\Psi_{rk}$. This, in turn, induces a partition of $V$ into at least two nonempty sets. Continuing in this way we construct a tree $T$ whose leaves are in 1-1 correspondence with the verifiers of $Rs$ and whose interior nodes correspond to the partitions needed to separate all the verifiers. Hence, there are at most $k$ leaves and at most $k-1$ partitions. Let $\Psi$ be the common refinement of all the partitions corresponding to the interior nodes of $T$. (By the common refinement of two partitions $\Psi_{ij}$ and $\Psi_{kr}$ we understand the partition $\{X \cap Y : X \in \Psi_{ij}, Y \in \Psi_{kr}\}$.) $\Psi$ has at most $m^{k-1}$ members.

STEP 2: Our next goal is to find another partition $\Phi^*$ of $\{p_0, \ldots, p_s\}$ such that

(a) for every $C$-consistent subset $V$ of $Ver_{Rs}$, the set $V^*$ obtained from $V$ by uniformly identifying variables belonging to the same member of $\Psi^*$ is $C$-consistent.

The construction of $\Psi^*$ goes as follows. For every $C$-consistent subset $V$ of $Ver_{Rs}$ there exists $h_V \in Hom(\mathcal{L}, M)$ and a set $d$ of designated truth-values of $M$ such that $h_V(V) \subseteq d$. We define two variables $p_i, p_j$ to be in the same member of the partition $\Psi_V$ if and only if $h_V(p_i) = h_V(p_j)$. Let $\Psi^*$ be the common refinement of all the partitions $\Psi_V$. Clearly, $\Psi^*$ satisfies (a): if $V$ and $V^*$ are as required, then $h_V(V^*) = h_V(V)$. Moreover, since $Ver_{Rs}$ has at most $2^k$ $C$-consistent subsets, $\Psi^*$ has at most $m^{2^k}$ members.

Now, we use the partitions $\Psi$ and $\Psi^*$, constructed in Steps 1 and 2, to find a substitution $e$ informally described at the beginning of the proof. First, let us note that the common refinement of $\Psi$ and $\Psi^*$ has at most $t$ members. Let $e$ be any substitution satisfying the following conditions:

(b) $e$ maps $\{p_0, \ldots, p_s\}$ into $\{p_0, \ldots, p_t\}$;

(c) for every $i, j \leq s$, $e(p_i) = e(p_j)$ *iff* $p_i$ and $p_j$ belong to the same element of the common refinement of $\Psi$ and $\Psi^*$.

In our final step of the proof, we show that the replacement of every verifier $v$ of $Rs$ with the formula $e(v)$, results in a minimal resolution counterpart $Rs^* = \langle \mathcal{V}^*, \mathcal{F}^* \rangle$ of $\mathcal{P}$. We begin with the proof of refutational completeness. By the definition of $\Psi$, $e(Ver_{Rs})$ has as many elements as $Ver_{Rs}$. For every $l$-ary connective $f$ of $\mathcal{L}$, the corresponding operation $f^*$ of $\mathcal{V}^*$ is defined by $f^*(e(v_1), \ldots, e(v_l)) = e(f_{Rs}(v_1, \ldots, v_l))$, where $f_{Rs}$ is the operation on $Ver_{Rs}$ corresponding to $f$. Hence, $e$ is an isomorphism between $\mathcal{V}$ and $\mathcal{V}^*$. In addition, we can easily verify that

$$\text{for every } V \subseteq Ver_{Rs}, e(V) \text{ is maximal consistent in } Rs^* \text{ iff } V \in \mathcal{D}.$$

In other words, $e$ is an isomorphism between the matrices induced by $Rs$ and $Rs^*$. Hence, by Proposition 1.8, $\mathcal{P}$ and $Rs^*$ satisfy (r1).

Next, let $f^*(e(v_1), \ldots, e(v_l)) \Rightarrow e(v)$ be a transformation rule of $Rs^*$. To show that

(r2) $f(e(v_1), \ldots, e(v_l)) \; \Theta_C \; e(v)$,

let us note that $f_{Rs}(v_1, \ldots, v_l) = v$. So, $f(v_1, \ldots, v_l) \; \Theta_C \; v$, from which we obtain (r2) using Proposition 1.5(ii).

Finally, to demonstrate (r3), let $V$ be a subset of $Ver_{Rs}$. If $C(e(V)) \neq L$, then, by Proposition 1.4, $C(V) \neq L$. This means that $V \notin \mathcal{F}$ and, hence, $e(V) \notin \mathcal{F}^*$. Conversely, suppose that $C(e(V)) = L$. By (a), $C(V) = L$ and $V \in \mathcal{F}$. So, by the definition of $\mathcal{F}^*$, $e(V) \in \mathcal{F}^*$. This shows (r3) and completes the proof of this theorem. ∎

Theorems 2.16, 2.17, and 2.18 describe some of the classes of resolution logics for which the minimization algorithm, supported by Theorem 4.6, will construct their minimal resolution counterparts. The restriction $\Theta_C^R = \Theta_C$ can be dropped from Theorem 4.6, if we decide to search for minimal strong resolution counterparts.

By Theorems 4.5 and 4.6, a minimal counterpart of a resolution logic $\mathcal{P}$ (that satisfies $\Theta_C^R = \Theta_C$) can be found by restricting our search to the sublanguage of $\mathcal{L}$ generated by $t = m^{2^k+k-1}$ variables ($m$ and $k$ are as in Theorem 4.6). Surprisingly, for many logics $t = 0$ suffices. For instance, a minimal resolution counterpart of classical propositional logic can be constructed from formulas containing just one variable. This shows that for some logics our bound $m^{2^k+k-1}$ on the number of variables necessary for the construction of a minimal resolution counterpart can be significantly improved.

## 4.2 VERIFIER DEGREES OF RESOLUTION LOGICS

As far as the computational complexity is concerned, we cannot expect much help from the minimization algorithm presented in the proof of Theorem 4.5 when its scope is the class of all resolution logics. It suffices to point out that the number of formulas in the sequence $\alpha_0, \ldots, \alpha_s$, which is constructed in the first step of the algorithm, can be equal to $m^{m^{t+1}}$ ($m$ and $t$ are as in the proof of Theorem 4.5). The search for minimal resolution counterparts can be made less stressful, if we restrict our attention to logical calculi for which $t = 0$, i.e., to resolution logics $\mathcal{P}$ such that

(v1) *there exists a minimal resolution counterpart $Rs$ of $\mathcal{P}$ such that $Ver_{Rs} \subseteq L^{(0)}$, i.e., $p_0$ is the only variable that may occur in the verifiers of $Rs$.*

Clearly, the satisfaction of (v1) does not necessarily imply that minimal resolution counterparts of a resolution logic are 'small'. In this section we investigate some of the classes of resolution logics which satisfy (v1) and discuss the computational complexity of the construction of the verifiers of their minimal resolution counterparts.

Let $\mathcal{P}$ be a resolution logic. The smallest number of different variables required to construct a minimal resolution counterpart of $\mathcal{P}$ is called the *verifier-degree* (*v-degree*) of $\mathcal{P}$. Hence, condition (v1) focuses our attention on resolution logics with $v$-degree $\leq 1$. By Theorem 4.3, for every $t > 0$ there exists a resolution logic of $v$-degree $\geq t$.

We may informally view $v$-degrees as a scale for classifying resolution logics with respect to the structure of verifiers of their minimal resolution counterparts or, looking at the minimization algorithm of Theorem 4.5, with respect to the complexity of the search for their minimal resolution counterparts. On the bottom of the scale there are logical calculi for which the number of variables occurring in their minimal resolution counterparts do not contribute significantly to the complexity of the minimization algorithm. Let us note that if a logic $\mathcal{P}$ is of $v$-degree 0, then among its minimal resolution counterparts there exists a resolution proof system whose verifiers are constructed using logical connectives and constants only. Słupecki's 3-valued logic, which will be discussed in Example 4.9, is of $v$-degree 0. Classical propositional logic may serve as an example of a logic of $v$-degree 1.

THEOREM 4.7: *Let* $M = \langle \mathcal{A}, \mathcal{D} \rangle$ *be a proper matrix for* $\mathcal{L}$ *such that* $\mathcal{A}$ *is definitionally complete and* $C_M(\emptyset) \neq \emptyset$*. Then v-degree of* $\langle \mathcal{L}, C_M \rangle$ *is* $\leq 1$*, and every minimal resolution counterpart of this logic has* $card(M)$ *verifiers.*

*Proof:* Let $M$ be as stated. Since $M$ is proper and finite (cf. [79], Section 4.4.1), by Theorem 3.7, $\mathcal{P} = \langle \mathcal{L}, C_M \rangle$ is a resolution logic.

Let $Rs = \langle \mathcal{V}, \mathcal{F} \rangle$ be a resolution proof system on $\mathcal{L}$ constructed as follows. For every truth-value $a$ of $M$, let $v_a$ be any formula that defines $a$. Clearly, $v_a$ can be selected from $L^{(0)}$ (indeed, if $a$ is defined by $\alpha_a$ containing more than one different variable, then the formula $v_a$ obtained from $\alpha_a$ by renaming all variables of $\alpha_a$ as $p_0$ also defines $a$). Let the algebra $\mathcal{V}$ be obtained from $\mathcal{A}$ by replacing every truth-value $a$ by $v_a$ and let $\mathcal{F}$ consist of all $C_M$-inconsistent sets of elements of $\mathcal{V}$.

We leave it to the reader to verify that $Rs$ is a resolution counterpart of $\mathcal{P}$. We only show that every resolution counterpart of $\mathcal{P}$ has at least as many verifiers as there are truth-values of $M$. To this end, let us suppose that $Rs' = \langle \mathcal{V}', \mathcal{F}' \rangle$ is an arbitrary resolution counterpart of $\mathcal{P}$ and let $M' = \langle \mathcal{V}', \mathcal{D}' \rangle$ be the matrix induced by $Rs'$. Let $d' \in \mathcal{D}'$ and let $N = \langle \mathcal{V}', \{d'\} \rangle / \Theta$, where $\Theta$ is the maximal congruence of $\langle \mathcal{V}', \{d'\} \rangle$. By Proposition 1.9 and Corollary 3.22, $C_N \geq C_M$. We claim that

(a) $card(N) \geq card(M)$.

Clearly, from (a) our theorem follows immediately. To prove (a) we first define a finite family $\mathcal{E}$ of formulas in two variables (say $p$ and $q$) such that

(b) for every $d \in \mathcal{D}$, the greatest (with respect to $\subseteq$) congruence $\Theta_d$ of the matrix $\langle \mathcal{A}, \{d\} \rangle$ satisfies the following equivalence

$$x \Theta_d\, y \textit{ iff } \text{ for every } \alpha(p,q) \in \mathcal{E}, h[p/x, q/y](\alpha(p,q)) \in d.$$

In (b), $h[p/x, q/y]$ is a valuation of $\mathcal{L}$ into $M$ such that $h[p/x, q/y](p) = x$ and $h[p/x, q/y](q) = y$. The definition of $\mathcal{E}$ is as follows. Let $a$ be an arbitrary element of $\bigcap \mathcal{D}$. Since $C_M(\emptyset) \neq \emptyset$, such an $a$ can always be selected. For every $d \in \mathcal{D}$, let $b_d$ be the truth-value of $M$ that is not in $d$, and let $f_d$ be the two-argument function on the base set $|\mathcal{A}|$ of $\mathcal{A}$ defined by

$$f_d(x, y) = \begin{cases} a, & \text{if } x = y, \\ b_d, & \text{otherwise.} \end{cases}$$

Since $\mathcal{A}$ is definitionally complete, for every $d \in \mathcal{D}$ there exists $\alpha_d$ that defines $f_d$. Put $\mathcal{E} = \{\alpha_d : d \in \mathcal{D}\}$. Then, for every $d \in \mathcal{D}$ and every $x, y \in |\mathcal{A}|$,

(c) $x = y$ *iff* for every $\alpha(p, q) \in \mathcal{E}, h[p/x, q/y](\alpha(p, q)) \in d$,

where $h[p/x, q/y]$ is as in (b). Now, to show that $\mathcal{E}$ satisfies (b), we only note that the definitional completeness of $\mathcal{A}$ implies that for every $d \in \mathcal{D}$, $\Theta_d$ is the identity relation (cf. [79], Section 4.4.1). Hence, (b) and (c) are equivalent.

Having (b) in our disposal, we apply the following fact proved by Czelakowski in [10] as Corollary 7.13 (see also [79], Theorem 3.4.7).

Lemma A: *Let* $M = \langle \mathcal{A}, \mathcal{D} \rangle$ *be a finite matrix satisfying* (b) *with respect to a finite family* $\mathcal{E}$ *of formulas in two variables. Moreover, let for every* $d \in \mathcal{D}$, $\langle \mathcal{A}, \{d\} \rangle$ *be factorial. Then every factorial matrix* $N$ *such that* $C_N \geq C_M$ *is isomorphic to a submatrix of a direct product of matrices of the form* $\langle \mathcal{A}, \{d\} \rangle$, *where* $d \in \mathcal{D}$.

By Lemma A, there exists a family $\{M_i : i \in I\}$ of matrices of the form $\langle \mathcal{A}, \{d\} \rangle$, $d \in \mathcal{D}$, such that the matrix $N$ (defined at the beginning of the proof) is isomorphic to a submatrix $N'$ of the product $\prod\{M_i : i \in I\}$. For simplicity, let us assume that $N = N'$. To conclude the proof of (a) it suffices to show that for every $a \in |\mathcal{A}|$, the tuple $\langle a_i : i \in I \rangle$ such that $a_i = a$, all $i \in I$, is a truth-value of $N$. To this end, let $h$ be any valuation of $\mathcal{L}$ into $N$, let $a \in |\mathcal{A}|$, and let $v_a$ be a formula defining $a$. Then, for every $i \in I$, there exists a valuation $h_i$ of $\mathcal{L}$ into $M_i$ such that for every formula $\beta \in L, h(\beta) = \langle h_i(\beta) : i \in I \rangle$. Hence, $h(v_a) = \langle h_i(v_a) : i \in I \rangle = \langle a_i : i \in I \rangle$, where $a_i = a$, all $i \in I$. This finishes the proof of (a) and the proof of our theorem has been completed. ∎

Using the terminology introduced in Section 3.5, Theorem 4.7 says that if a resolution class $\mathcal{K}_\mathcal{P}$ of $\mathcal{P}$ contains a definitionally complete logic, then among the minimal resolution counterparts of $\mathcal{P}$ there is a resolution proof system whose verifiers are constructed from logical connectives, constants, and at most one propositional variable.

Our next theorem has a similar meaning. If $M$ is a matrix for $\mathcal{L}$, then by $M^*$ let us denote the quotient of $M$ modulo the greatest congruence of $M$ (with respect to $\subseteq$).

THEOREM 4.8: *Let $\mathcal{P}$ be an implicational logic defined by a finite matrix $M = \langle \mathcal{A}, \{d\} \rangle$ whose every truth-value is definable. Then $\mathcal{P}$ is of v-degree $\leq 1$ and every minimal resolution counterpart of $\mathcal{P}$ has $card(M^*)$ verifiers.*

*Proof:* Let $\mathcal{P}$ and $M = \langle \mathcal{A}, \{d\} \rangle$ be as stated. Let us recall from the proof of Theorem 2.18 that every truth-value of $M^*$ is definable. Hence, without any loss of generality we can assume that $M = M^*$.

As in the proof of Theorem 4.7, we justify the existence of a resolution counterpart of $\mathcal{P}$ with $card(M)$ verifiers, all of which are formulas of $L^{(0)}$. To conclude the proof we must show that every resolution counterpart of $\mathcal{P}$ has at least $card(M)$ verifiers. Let $Rs = \langle \mathcal{V}, \mathcal{F} \rangle$ be an arbitrary resolution counterpart of $\mathcal{P}$ and let $d_N$ be one of the maximal consistent sets of verifiers of $Rs$. Let $N$ be $\langle \mathcal{V}, \{d_N\} \rangle^*$. We claim that

(a) $card(N) \geq card(M)$.

To prove the claim, let us note that $M$ satisfies the condition (b) formulated in the proof of Theorem 4.7 with respect to the family $\mathcal{E} = \{p \to q, q \to p\}$ (cf. [10], Proposition 8.1). Moreover, by Proposition 1.9 and Corollary 3.22, $C_N \geq C_M$. By Lemma A, quoted in the proof of Theorem 4.7, we may therefore assume that $N$ is a submatrix of the direct product of some number of copies of $M$. Now, we can repeat the argument presented at the end of the proof of Theorem 4.7 to show that for every $a \in |\mathcal{A}|$, $\langle a, a, \ldots \rangle$ is a truth-value of $N$. This shows (a) and the proof of our theorem is finished. ∎

If a logic $\mathcal{P}$ is defined by a matrix $M$ which satisfies either Theorem 4.7 or 4.8, then one way of constructing a minimal resolution counterpart $Rs = \langle \mathcal{V}, \mathcal{F} \rangle$ of $\mathcal{P}$ is to run the minimization algorithm of Theorem 4.5 with $t = 0$. Another, more efficient way is to form $\mathcal{V}$ from the algebra of truth-values of $M$ by replacing every truth-value of $M$ with a formula that defines it, and by taking $\mathcal{F}$ to be the family of all inconsistent sets of verifiers of $Rs$. Clearly, in this case, the problem of the construction of the verifiers of $Rs$ reduces to the problem of the search for formulas defining the truth-values of $M$.

EXAMPLE 4.9: Theorems 4.7 and 4.8 provide us with other tools that can be used to demonstrate that the system $Rs_2$ defined in Example 3.15 is a minimal resolution counterpart of classical propositional logic, and that this logic is of v-degree 1.

The so-called Słupecki logic $\mathcal{P}_s$ is an example of a logical calculus of $v$-degree 0. $\mathcal{P}_s = \langle \mathcal{L}_s, C_s \rangle$ is defined by the 3-valued matrix $M_s = \langle \mathcal{A}_s, \{2\} \rangle$, where $\mathcal{A}_s$ is the algebra $\langle \{0, 1, 2\}, \underline{\neg}, \underline{\vee}, \underline{\wedge}, \underline{\rightarrow}, \underline{1} \rangle$ and the operations of $\mathcal{A}_s$ are defined in the same way as in the 3-valued Łukasiewicz matrix (cf. Example 1.6). $\underline{1}$ is the constant function.

$\mathcal{L}_s$ has all the connectives of the language of classical logic plus the logical constant **1** interpreted as the constant function $\underline{1}$. The logic $\mathcal{P}_s$, which is also known as Słupecki's variant of the 3-valued Łukasiewicz logic (cf. [5, 53]), is definitionally complete and implicational. Hence, by either Theorem 4.7 or Theorem 4.8, it is of $v$-degree $\leq 1$. Since the verifiers of the minimal resolution counterpart $Rs_3^s$ of $\mathcal{P}_s$ (cf. Appendix F) do not contain variables, this logic is of $v$-degree 0. ∎

## 4.3 MINIMAL RESOLUTION COUNTERPARTS OF ŁUKASIEWICZ LOGICS

Among logical calculi that satisfy the assumptions of neither Theorem 4.7 nor Theorem 4.8 are finitely-valued Łukasiewicz logics. In this section we show that every minimal resolution counterpart of the $n$-valued Łukasiewicz logic $\mathcal{P}_n$ has $2n$ verifiers (with the exception of $n = 2$; every minimal resolution counterpart of $\mathcal{P}_2$ has only two verifiers, cf. Corollary 4.2).

Let $\mathcal{L}$ be a propositional language with a binary connective $\vee$. We say that a matrix $M = \langle \mathcal{A}, \mathcal{D} \rangle$ for $\mathcal{L}$ is *well-connected* (with respect to $\vee$) if and only if for every $d \in \mathcal{D}$ and every truth-values $a, b$ of $M$,

$$a \vee b \in d \textit{ iff } a \in d \text{ or } b \in d.$$

(In this definition we use '$\vee$' to denote both a logical connective and the operation of $M$ corresponding to this connective.) It is an easy exercise to verify that the logic defined by a well-connected matrix is disjunctive. By Lemma 3.10, the matrix induced by a resolution counterpart of a disjunctive logic is always well-connected.

LEMMA 4.10 (Harley): *Let $M = \langle \mathcal{A}, \{d\} \rangle$ be a finite and well-connected matrix with $n$ truth-values, and let $Rs$ be a resolution proof system on $\mathcal{L}$ satisfying* (r2) *and* (r3) *with respect to $C_M$. If there is $h \in Hom(\mathcal{L}, M)$ that maps $Ver$ onto the set of all truth-values of $M$, then $Rs$ is a resolution counterpart of $\langle \mathcal{L}, C_M \rangle$. Moreover, the resolution rule of $Rs$ can be simplified to the form*

$$Res_n: \frac{\alpha_0(p), \ldots, \alpha_{n-1}(p)}{\alpha_0(p/v_0) \vee \ldots \vee \alpha_{n-1}(p/v_{n-1})},$$

*where $v_0, \ldots, v_{n-1}$ is an arbitrary but fixed list of verifiers of $Rs$ that are mapped by $h$ onto the set of truth-values of $M$.*

*Proof:* Let $Rs = \langle \mathcal{V}, \mathcal{F} \rangle, M$, and $h$ be as stated. For reasons of simplicity, let us assume that the truth-values of $M$ are the integers $0, 1, \ldots, n-1$. We prove only the second part of the lemma, leaving the proof of the first part to the reader.

Let $Ver = \{v_0, \ldots, v_{s-1}\}$ and let us assume that $v_0, \ldots, v_{n-1}$ is a list of verifiers of $Rs$ such that for every $i < n, h(v_i) = i$. We want to show that the replacement of the resolution rule of $Rs$ with the rule $Res_n$ implies the refutational completeness condition (r1).

To this end, let $X \subseteq L$ be finite and clean in $Rs$. First, let us assume that $C_M(X) = L$. Since $Rs$ is a counterpart of $\langle \mathcal{L}, C_M \rangle$, the inference rules of $Rs$ can be used to deduce verifiers $w_0, \ldots, w_k$ of $Rs$ such that $\{w_0, \ldots, w_k\} \in \mathcal{F}$. We claim that using the rule $Res_n$ and the transformation rules of $Rs$, we can deduce verifiers $w_0^*, \ldots, w_k^*$ such that $\{w_0^*, \ldots, w_k^*\} \in \mathcal{F}$. To prove this claim, let us consider a verifier $w_i, i \leq k$. Without any loss of generality we can assume that $w_i$ has been deduced from $X$ in two phases: in phase one the formula $\alpha_i$ has been deduced using only the resolution rule, and, in phase two, the transformation rules of $Rs$ have been applied to reduce $\alpha_i$ to $w_i$. By Lemma 3.17,

$$\alpha_i = \bigvee_{1 \leq j \leq s^t} \beta_j(p_1/v_{\rho_j(1)}, \ldots, p_t/v_{\rho_j(t)}),$$

where $\rho_1, \ldots, \rho_{s^t}$ are all the mappings from $\{1, \ldots, t\}$ into $\{0, \ldots, s-1\}$, and every $\beta_j$ is in $X$. (We assume here that $X \subseteq L^{(t)}$ and, to simplify this proof, that $Var(Ver) = \{p_0\}$.) Let $\alpha_i^*$ be obtained from $\alpha_i$ by removing any disjunct that contains, as a subformula, a verifier from $\{v_n, \ldots, v_{s-1}\}$. Let us note that $\alpha_i^*$ can be deduced from $X$ by applying $Res_n$ the same number of times as it was required to deduce $\alpha_i$ from $X$ using the resolution rule of $Rs$. We claim that

(a) $C_M(\alpha_0^*, \ldots, \alpha_k^*) = L$.

Indeed, if for some $h^* \in Hom(\mathcal{L}, M), h^*(\{\alpha_0^*, \ldots, \alpha_k^*\}) \subseteq d$, then, by well-connectivity of $M, h^*(\{\alpha_0, \ldots, \alpha_k\}) \subseteq d$. This would mean that $C_M(\alpha_0, \ldots, \alpha_k) \neq L$ and, by (r2) and (r3), that $\{w_0, \ldots, w_k\} \notin \mathcal{F}$ – contradiction. By (a), (r2), and (r3), we can apply the transformation rules to $\{\alpha_0^*, \ldots, \alpha_k^*\}$ to deduce $\{w_0^*, \ldots, w_k^*\} \in \mathcal{F}$. This concludes the first part of the proof of (r1).

For the second part, let us assume that $X$ is $C_M$-consistent but refutable in $Rs$ using $Res_n$. This means that for some valuation $h$ of $\mathcal{L}$ into $M, h(X) \subseteq d$. Moreover, there exists a sequence

(b) $X = X_0 \subseteq X_1 \subseteq \ldots \subseteq X_k$

of finite sets of formulas such that: $X_k$ contains a set $Y$ from $\mathcal{F}$, and for every $i < k$, $X_{i+1}$ is obtained from $X_i$ by the application of $Res_n$ or a transformation rule of $Rs$. In other words, the sequence (b) represents the steps in the refutation of $X$. Without any loss of generality, we can assume that in (b) no application of a transformation rule precedes an application of $Res_n$. We claim that these assumptions lead to a contradiction.

Let $X_i$ be the first set in (b) to which a transformation rule has been applied. Then

(c) for every $l \leq i, h(X_l) \subseteq d$.

Clearly, (c) holds for $l = 0$. Let us assume that (c) holds for some $l < i$, and that $\alpha_0(p), \ldots, \alpha_{n-1}(p) \in X_l$ are such that $X_{l+1} = X_l \cup \{\beta\}$, where $\beta = \alpha_0(p/v_0) \vee \ldots \vee \alpha_{n-1}(p/v_{n-1})$. If $h(p) = j$, then $h(\alpha_j(p/v_j)) = h(\alpha_j(p)) \in d$, and, since $M$ is well-connected, $h(\beta) \in d$. This shows that $h(X_{l+1}) \subseteq d$ and the proof of (c) has been finished. By (c), $X_i$ is $C_M$-consistent and, by (r2), so are the remaining sets $X_{i+1}, \ldots, X_k$. This would mean, however, that $Y$ is consistent which, by (r3), is impossible. Therefore, if $X$ is refutable using $Res_n$, then $X$ is $C_M$-inconsistent. ∎

Lemma 4.10 suggests that the resolution rule of some resolution counterparts can be substantially simplified. We shall explore this idea in the next section. For now, our objective is to use Lemma 4.10 as a test for recognizing resolution counterparts of Łukasiewicz logics.

Let $\mathcal{P}_n = \langle \mathcal{L}, C_n \rangle$ be an arbitrary but fixed Łukasiewicz logic, $n > 2$, and let $Rs_n = \langle \mathcal{V}_n, \mathcal{F}_n \rangle$ be a resolution proof system on $\mathcal{L}$ defined as follows. For every $i < n$, let $v_i(p_0)$ be a formula such that for every $h \in Hom(\mathcal{L}, M_n)$,

$$h(v_i(p_0)) = \begin{cases} i, & \text{if } h(p_0) = n-2, \\ 0, & \text{otherwise.} \end{cases}$$

The existence of such formulas is guaranteed by the McNaughton test of Proposition 1.7. For every $n \leq i < 2n$, let $v_i = \neg v_{2n-1-i}$. Let

$$Ver = \{v_i(p_0) : i < 2n\}$$

be the set of verifiers of $Rs_n$ and let

$$\mathcal{V}_n = \langle Ver, \neg, \vee, \wedge, \rightarrow \rangle.$$

The operations $\neg, \vee, \wedge, \rightarrow$ of the algebra $\mathcal{V}_n$ (which correspond to the logical connectives of $\mathcal{L}$ of the same names) are defined as follows: for every $v_i, v_j, v_k \in Ver$,

$\neg v_i = v_{2n-1-i}$ ;

$v_i \vee v_j = v_k$ *iff* $h(v_k) = max(h(v_i), h(v_j))$, for every $h \in Hom(\mathcal{L}, M_n)$;

$v_i \wedge v_j = v_k$ *iff* $h(v_k) = min(h(v_i), h(v_j))$, for every $h \in Hom(\mathcal{L}, M_n)$;

$v_i \rightarrow v_j = v_k$ *iff* $h(v_k) = min(n-1, n-1-h(v_i)+h(v_j))$, for every $h \in Hom(\mathcal{L}, M_n)$.

Finally, the family $\mathcal{F}_n$ consists of all $C_n$-inconsistent subsets of $Ver$.

TABLE 4.1: Transformation rules of $Rs_5$ for $\rightarrow$.

| $\rightarrow$ | 0 | 1 | 2 | 3 | 4 | 5 | 6 | 7 | 8 | 9 |
|---|---|---|---|---|---|---|---|---|---|---|
| 0 | 9 | 9 | 9 | 9 | 9 | 9 | 9 | 9 | 9 | 9 |
| 1 | 8 | 9 | 9 | 9 | 9 | 8 | 9 | 9 | 9 | 9 |
| 2 | 7 | 8 | 9 | 9 | 9 | 7 | 8 | 9 | 9 | 9 |
| 3 | 6 | 7 | 8 | 9 | 9 | 6 | 7 | 8 | 9 | 9 |
| 4 | 5 | 6 | 7 | 8 | 9 | 5 | 6 | 7 | 8 | 9 |
| 5 | 4 | 4 | 4 | 4 | 4 | 9 | 9 | 9 | 9 | 9 |
| 6 | 3 | 4 | 4 | 4 | 4 | 8 | 9 | 9 | 9 | 9 |
| 7 | 2 | 3 | 4 | 4 | 4 | 7 | 8 | 9 | 9 | 9 |
| 8 | 1 | 2 | 3 | 4 | 4 | 6 | 7 | 8 | 9 | 9 |
| 9 | 0 | 1 | 2 | 3 | 4 | 5 | 6 | 7 | 8 | 9 |

EXAMPLE 4.11: For $n = 5$ the system $Rs_n$ described above has the following verifiers:

$v_0 = \neg(p_0 \to p_0)$, $v_1 = \neg(H_4 \to H_3)$, $v_8 = \neg v_1$,
$v_2 = v_8 \to v_1$, $v_3 = v_8 \to v_2$, $v_4 = v_8 \to v_3$,
$v_5 = \neg v_4$, $v_6 = \neg v_3$, $v_7 = \neg v_2$,
$v_9 = \neg v_0$,

where $H_3 = p_0 \to (p_0 \to \neg p_0)$ and $H_4 = p_0 \to H_3$.

Table 4.1 lists the transformation rules of $Rs_5$ for the implication connective. The transformation rules for the remaining connectives can be found in Appendix D. ∎

THEOREM 4.12 (Harley, Stachniak): *$Rs_n$ is a resolution counterpart of $\mathcal{P}_n$.*

*Proof:* Clearly, $Rs_n$ satisfies (r2) and (r3). Let $h \in Hom(\mathcal{L}, M_n)$ be such that $h(p_0) = n-2$. Then $h(Ver) = \{0, \ldots, n-1\}$ and our theorem follows from Lemma 4.10. ∎

So far we have shown that $Rs_n$ is a resolution counterpart of $\mathcal{P}_n$. To show that $Rs_n$ is minimal among the resolution counterparts of $\mathcal{P}_n$ (and, hence, that $v$-degree of $\mathcal{P}_n$ is 1) we need the following lemma.

LEMMA 4.13 (Harley): *If $Rs$ is a resolution counterpart of $\mathcal{P}_n$, then there exists $h \in Hom(\mathcal{L}, M_n)$ that maps the set of verifiers of $Rs$ onto the set of truth-values of $M_n$.*

*Proof:* Let Rs be a resolution counterpart of $\mathcal{P}_n$, with $Ver = \{v_0, \ldots, v_k\}$ as its set of verifiers, and let $X = \{\alpha_i(p_i) : i < n\} \subseteq L$ be selected in the following way. $p_i$ is the only variable occurring in $\alpha_i$ and this variable occurs neither in verifiers of $Rs$ nor the remaining formulas of $X$. Moreover, for every $\alpha_i(p_i) \in X$ and every $h \in Hom(\mathcal{L}, M_n), h(\alpha_i) = n-1$, if $h(p_i) = i$, and $h(\alpha_i) = 0$, in the remaining cases. Such formulas exist by the McNaughton test. Clearly, the set $X$ is consistent. If no valuation satisfies the conclusion of the lemma, i.e., if

(a) $h(Ver) \neq \{0, \ldots, n-1\}$, for every $h \in Hom(\mathcal{L}, M_n)$,

then $X$ would be refutable in $Rs$, which, in the light of (r1) and the consistency of $X$, is impossible. Indeed, if (a) were true, then for every $i < n$, we could form the resolvent $\beta_i = \alpha_i(p_i/v_0) \vee \ldots \vee \alpha_i(p_i/v_k)$. By (a), $\{\beta_i : i \leq n-1\}$ would be inconsistent and, by (r2) and (r3), it would be reducible to an inconsistent subset of $Ver$. Hence, $X$ would be refutable. ∎

THEOREM 4.14 (Harley, Stachniak): *Any resolution counterpart of $\mathcal{P}_n$ has at least $2n$ verifiers.*

*Proof:* Let $Rs$ be a resolution counterpart of $\mathcal{P}_n$ with $Ver$ as its set of verifiers. Let $h_0, h_1 \in Hom(\mathcal{L}, M_n)$ be such that $h_0(p) = 0$, for every variable $p$, and $h_1(Ver) = \{0, \ldots, n-1\}$ (cf. Lemma 4.13). For every $v \in Ver$, let $H(v) = \langle h_0(v), h_1(v) \rangle$. By the McNaughton test, $h_0(Ver) \subseteq \{0, n-1\}$. We leave it to the reader to verify that for every $i < n$, there are $v, w \in Ver$ such that $H(v) = \langle 0, i \rangle$ and $H(w) = \langle n-1, i \rangle$. ∎

From Theorem 4.14 we immediately conclude that $Rs_n$ (as constructed in this section) is a minimal resolution counterpart of $\mathcal{P}_n$ and that:

COROLLARY 4.15 (Harley, Stachniak): *For every $n > 2$, $\mathcal{P}_n$ is of v-degree 1, and every minimal resolution counterpart of $\mathcal{P}_n$ has $2n$ verifiers.*

In applications where the syntactic structure of verifiers of a given resolution proof system is of no significance, the verifiers can be represented in the resolution rule as constants. When such syntactic descriptions are essential, we need an algorithm to print them out. We conclude this section by giving a polynomial time algorithm for generating the verifiers of $Rs_n$, for every $n > 2$. Let $\mathcal{P}_n, n > 2$, be fixed for the remainder of this section. For every $t \geq 1$, let $H_t(p_0)$ be the formula defined as follows:

$$H_1(p_0) = \neg p_0, \quad H_{t+1}(p_0) = p_0 \to H_t(p_0).$$

The truth function $h_t(x)$ defined by $H_t(p_0)$, $t \geq 1$, satisfies

$$h_t(x) = min(n-1, t \cdot (n-1-x))$$

(cf. [55]). It can be easily verified that the truth function $g(x)$ defined by the formula $v_1 = \neg(H_{n-1}(p_0) \to H_{n-2}(p_0))$ satisfies

$$g(x) = \begin{cases} 0, & \text{if } x \neq n-2, \\ 1, & \text{if } x = n-2, \end{cases}$$

and, hence, $v_1$ is the verifier semantically described in the paragraph following Lemma 4.10. The remaining verifiers can be generated in the following way. We define:

$$v_0 = \neg(p_0 \to p_0), \qquad v_{2n-1} = p_0 \to p_0, \qquad v_{2n-2} = \neg v_1,$$

and for $2 \leq i < n$,

$$v_i = v_{2n-2} \to v_{i-1}, \qquad v_{2n-1-i} = \neg v_i.$$

Both $H_{n-1}(p_0)$ and $H_{n-2}(p_0)$ can be constructed in $n$ steps. Hence, $v_1$ and $v_{2n-2}$ can be printed out in $O(n)$ steps. For every $2 \leq i < n$, the verifiers $v_i$ and $v_{2n-1-i}$ require up to $O(n^2)$ steps to be printed out. To conclude, the verifiers of $Rs_n$ can be printed out in $O(n^3)$ steps.

## 4.4 SIMPLIFICATION OF THE RESOLUTION RULE

Lemma 4.10 describes one class of logics whose resolution counterparts can use a simpler form of the resolution rule, i.e, the resolution rule of these resolution proof systems can be written using a proper subset of the set of verifiers. The replacement of the resolution rule $R^{\vee}_{Ver}$ by its simplified form $Res_n$ reduces not only the size of the search space but also the length of resolvents, and, hence, speeds up the deductive process by eliminating many applications of the inference rules. Indeed, an examination of the proof of Lemma 4.10 reveals that every refutation can be simulated by a shorter deduction when the replacement of $R^{\vee}_{Ver}$ by $Res_n$ is allowed. Let us illustrate these remarks with the following example.

EXAMPLE 4.16: By Lemmas 4.10 and 4.13, the resolution rule of the minimal resolution counterpart $Rs_n$ of $\mathcal{P}_n$, $n > 2$, can be written in terms of $n$ verifiers instead of all $2n$. Let us look at the system $Rs_5$ described in Example 4.11 (see also Appendix D). Consider the set

$$X = \{p_1, p_1 \to p_2, p_2 \to p_3, \neg p_3\}.$$

$X$ is inconsistent in $\mathcal{P}_5$ and its refutation in $Rs_5$ can be initiated in one of 3072 ways, as this is the number of possible applications of the resolution rule of $Rs_5$ to the formulas of $X$. However, if we replace the resolution rule of $Rs_5$ by the restricted resolution rule $Res_5$, which requires only five verifiers, then there are only 96 possible ways to start a refutation of $X$.

Let us assume that $Res_5$ is defined in terms of the verifiers $v_0, \ldots, v_4$ (clearly, if $h \in Hom(\mathcal{L}, M_5)$ is such that $h(p_0) = 3$, then $h(\{v_0, \ldots, v_4\}) = \{0, \ldots, 4\}$). We can begin a refutation of $X$ by applying $Res_5$ to the sequence

$$p_2 \to p_3,\ \neg p_3,\ \neg p_3,\ \neg p_3,\ \neg p_3,$$

resulting in the resolvent $(p_2 \to v_0) \vee \neg v_1 \vee \neg v_2 \vee \neg v_3 \vee \neg v_4$, which reduces to

$$\beta_1(p_2) = (p_2 \to v_0) \vee v_8.$$

The second application of $Res_5$ uses the formulas:

$$p_1 \to p_2,\ p_1 \to p_2,\ p_1 \to p_2,\ p_1 \to p_2, \beta_1(p_2)$$

to generate the resolvent

$$(p_1 \to v_0) \vee (p_1 \to v_1) \vee (p_1 \to v_2) \vee (p_1 \to v_3) \vee \beta_1(p_2/v_4),$$

which reduces to

$$\beta_2(p_1) = (p_1 \to v_0) \vee (p_1 \to v_1) \vee (p_1 \to v_2) \vee (p_1 \to v_3) \vee v_8.$$

For the third application of $Res_5$ we use the sequence

$$p_1,\ p_1,\ p_1,\ p_1,\ \beta_2(p_1)$$

to obtain $v_0 \vee v_1 \vee v_2 \vee v_3 \vee \beta_2(p_1/v_4)$, which reduces to $v_8$. Finally, we apply the resolution rule to the sequence

$$p_1,\ p_1,\ p_1,\ p_1,\ p_1,$$

and obtain $v_0 \vee v_1 \vee v_2 \vee v_3 \vee v_4$, which reduces to $v_4$. Since $\{v_4, v_8\}$ is inconsistent in $\mathcal{P}_5$, the deductive process terminates successfully and, by (r1), the inconsistency of $X$ has been confirmed. ∎

The range of application of Lemma 4.10 goes beyond the class of finitely-valued Łukasiewicz logics. Our next theorem describes a larger class of resolution logics which satisfy the assumptions of this lemma. We begin by introducing the following notation. If $M$ is a matrix for $\mathcal{L}$, then by $\Theta_M$ we denote the congruence of $\mathcal{L}$ defined as follows: for every $\alpha, \beta \in L$,

$$\alpha \Theta_M \beta \text{ iff for every } h \in Hom(\mathcal{L}, M), h(\alpha) = h(\beta).$$

THEOREM 4.17: *Let $M = \langle \mathcal{A}, \{d_M\} \rangle$ be a finite, proper, and well-connected matrix such that:*

(i) *for every proper submatrix $N$ of $M$, there is a set $X \subseteq L$ such that $C_N(X) = L$ and $C_M(X) \neq L$;*

(ii) $\Theta^R_{C_M} = \Theta_M$.

*Then, for every resolution counterpart $Rs$ of $\langle \mathcal{L}, C_M \rangle$ there is $h \in Hom(\mathcal{L}, M)$ that maps $Ver_{Rs}$ onto the set of truth-values of $M$.*

Before we prove this theorem, let us explain the role of condition (i). If $M$ and its certain proper submatrix $N$ violate condition (i), then, informally speaking, $M$ is too 'big' for our purposes and should be replaced by a smaller matrix such as $N$. Indeed, in such a case $N$ would have fewer truth-values than $M$, would be well-connected, and would define the logic which had the same inconsistent sets as the logic defined by $M$. So, we would have $\Theta^R_{C_N} = \Theta^R_{C_M} = \Theta_M \subseteq \Theta_N$. Since $\Theta_N \subseteq \Theta^R_{C_N}$, we would also have $\Theta_N = \Theta^R_{C_N}$. (See also Exercise 4.6.)

The proof of Theorem 4.17 is based on a number of lemmas presented below. If $Rs$ is a resolution proof system on $\mathcal{L}$ and if $X \subseteq L$, then by $r^*(X)$ we denote the set of all verifiers of $Rs$ that can be deduced from $X$ using the inference rules of $Rs$.

LEMMA 4.18: *Let $M = \langle \mathcal{A}, \{d_M\} \rangle$ be a finite, proper, and well-connected matrix for $\mathcal{L}$ and let $Rs$ be a resolution counterpart of $\langle \mathcal{L}, C_M \rangle$. There exists $h_{Ver} \in Hom(\mathcal{L}, M)$ such that for every finite, clean in $Rs$, and $C_M$-consistent set $X$, $h_{Ver}(r^*(X)) \subseteq d_M$. Moreover, $h^{-1}_{Ver}(d_M) \cap Ver$ is maximal consistent in $Rs$.*

*Proof:* Let $M$ and $Rs$ be as stated, and let us assume that no valuation of $\mathcal{L}$ into $M$ satisfies the conclusion of the lemma. In other words, for every valuation $h_i \in Hom(\mathcal{L}, M)$, there is a finite, clean in $Rs$, and $C_M$-consistent set $X_i$ such that $h_i(r^*(X_i)) \not\subseteq d_M$. Since $M$ and $Var(Ver)$ are finite, we can restrict our attention to a finite list $h_0, \ldots, h_s$ of valuations representing all possible assignments of truth-values of $M$ to the variables of $Var(Ver)$. Moreover, without any loss of generality, we can assume that for every $i, j \leq s, i \neq j$ implies $Var(X_i) \cap Var(X_j) = \emptyset$.

Let $X = X_0 \cup \ldots \cup X_s$. Clearly, $X$ is $C_M$-consistent. By Theorem 3.18 and (r2), $r^*(X)$ is $C_M$-consistent, that is, for some $h \in Hom(\mathcal{L}, M), h(r^*(X)) \subseteq d_M$. Let

$i \leq s$ be such that for every $p \in Var(Ver), h_i(p) = h(p)$. Clearly, $h_i(r^*(X)) \subseteq d_M$, and, since $r^*(X_i) \subseteq r^*(X)$, we have $h_i(r^*(X_i)) \subseteq d_M$. This contradicts the choice of $X_i$. Hence, there exists $h \in Hom(\mathcal{L}, M)$ such that for every finite, clean in $Rs$, and $C_M$-consistent set $X, h(r^*(X)) \subseteq d_M$.

Let $d^*$ be a maximal consistent set of verifiers such that $h^{-1}(d_M) \cap Ver \subseteq d^*$ and let $h_{Ver} \in Hom(\mathcal{L}, M)$ be such that $h_{Ver}(d^*) \subseteq d_M$. Then, for every finite, clean in $Rs$, and $C_M$-consistent set $X, r^*(X) \subseteq d^*$ and we conclude that $h_{Ver}(r^*(X)) \subseteq d_M$. ∎

LEMMA 4.19: *Let $M$ be as in Theorem 4.17 and let $Rs$ be a resolution counterpart of $\langle \mathcal{L}, C_M \rangle$. Moreover, let $h_{Ver}$ be any valuation satisfying Lemma 4.18. Then $h_{Ver}$ defines a matrix homomorphism of $N = \langle \mathcal{V}, \{h_{Ver}^{-1}(d_M) \cap Ver\} \rangle$ into $M$. Moreover, the logics defined by $M$ and $N$ have the same inconsistent sets.*

*Proof:* Let $M, Rs, h_{Ver}$, and $N$ be as stated, and let $d^* = h_{Ver}^{-1}(d_M) \cap Ver$. By Lemma 4.18, $d^*$ is maximal consistent in $Rs$. To show that $h_{Ver}$ defines a homomorphism of $N$ into $M$, let us note that if $f$ is a $k$-ary connective of $\mathcal{L}$ and $f_{Rs}$ and $f_M$ are the operations of $\mathcal{V}$ and $\mathcal{A}$ corresponding to $f$, then for every $v_1, \ldots, v_k \in Ver, f_{Rs}(v_1, \ldots, v_k)\ \Theta_{C_M}^R f(v_1, \ldots, v_k)$. By Theorem 4.17(ii), this means that $h_{Ver}(f_{Rs}(v_1, \ldots, v_k)) = h_{Ver}(f(v_1, \ldots, v_k)) = f_M(h_{Ver}(v_1), \ldots, h_{Ver}(v_k))$.

To show that the logics defined by $M$ and $N$ have the same inconsistent sets, let us note that, by Theorem 3.5, for every $X \subseteq L, C_M(X) = L$ implies $C_N(X) = L$. Now, let us assume that for some $X \subseteq L, C_N(X) = L$. Without any loss of generality we can assume that $X$ is finite and clean in $Rs$ (recall that $\langle \mathcal{L}, C_N \rangle$ is compact and there is a finite $C_N$-inconsistent set). Let $Var(X) = \{p_1, \ldots, p_s\}$ and let $Ver = \{v_0, \ldots, v_{n-1}\}$. For every $h \in Hom(\mathcal{L}, N)$ there is $\alpha_h \in X$ such that $h(\alpha_h) \notin d^*$. In other words, for every mapping $\rho$ from $\{1, \ldots, s\}$ into $\{0, \ldots, n-1\}$ there is $\alpha_\rho \in X$ such that $\alpha_\rho(p_1/v_{\rho(1)}, \ldots, p_s/v_{\rho(s)})$ can be transformed, using the transformation rules of $Rs$, into a verifier not in $d^*$. From Lemma 3.17 it follows that using the resolution rule of $Rs$ we can form the resolvent

$$\alpha = \bigvee_{1 \leq i \leq n^s} \alpha_{\rho_i}(p_1/v_{\rho_i(1)}, \ldots, p_s/v_{\rho_i(s)}),$$

where $\rho_1, \ldots, \rho_{n^s}$ are all the mappings from $\{1, \ldots, s\}$ into $\{0, \ldots, n-1\}$. (For instance, if $s = n = 2$, then $\alpha = \alpha_1(p_1/v_0, p_2/v_0) \vee \alpha_2(p_1/v_1, p_2/v_0) \vee \alpha_3(p_1/v_0, p_2/v_1) \vee \alpha_4(p_1/v_1, p_2/v_1)$.) Moreover, the formulas $\alpha_{\rho_i}$ can be selected so that every $\alpha_{\rho_i}(p_1/v_{\rho_i(1)}, \ldots, p_s/v_{\rho_i(s)})$ is reducible to a verifier not in $d^*$. By Lemma 3.10, this means that $\alpha$ can be transformed into a verifier $v_\alpha \notin d^*$. Since $v_\alpha \in r^*(X)$, by Lemma 4.18 and the definition of $d^*$, we must have $C_M(X) = L$. ∎

From Lemma 4.19, Theorem 4.17 follows immediately. Indeed, if $Rs$ were a resolution counterpart of $\langle \mathcal{L}, C_M \rangle$ such that the valuation $h_{\mathcal{V}_{er}}$ described in Lemma 4.18 were not 'onto', then the $h_{\mathcal{V}_{er}}$ image of the matrix $N$ described in Lemma 4.19 would be a proper submatrix of $M$ and the condition (i) of Theorem 4.17 would be violated.

In the light of Lemma 4.10, Theorem 4.17 describes a class of resolution logics whose resolution counterparts can use the simpler form $Res_n$ of the resolution rule. Since every Łukasiewicz matrix $M_n$ satisfies the assumptions of Theorem 4.17 (cf. Exercise 2.7), this theorem restates the result reported in Lemma 4.13.

Another class of resolution logics to which Theorem 4.17 applies is the class of Post logics. Indeed, let $C_n^*$ denote the consequence operation defined by the $n$-valued Post matrix $M_n^*$. Since $M_n^*$ does not have non-trivial submatrices, condition (i) of Theorem 4.17 is satisfied. To show Theorem 4.17(ii), let us note that the inclusion $\Theta_M \subseteq \Theta_{C_M}^R$ holds for every proper matrix $M$. Next, if for some formulas $\alpha$ and $\beta$ there exists a valuation $h$ of $\mathcal{L}$ into $M_n^*$ such that $h(\alpha) \neq h(\beta)$, then we can select an integer $k$ such that $\neg^k\beta \notin C_n^*(\neg^k\alpha)$. This shows that $\alpha$ and $\beta$ are not $\Theta_{C_n^*}$-congruent. By Theorem 2.18, $\alpha$ and $\beta$ cannot be $\Theta_{C_n^*}^R$-congruent as well. This demonstrates Theorem 4.17(ii).

## 4.5 SIMPLIFICATION OF THE TERMINATION TEST

Let $\mathcal{P} = \langle \mathcal{L}, C \rangle$ be a disjunctive resolution logic and let $Rs = \langle \mathcal{V}, \mathcal{F} \rangle$ be a resolution counterpart of $\mathcal{P}$. As explained in Section 2.3, the family $\mathcal{F}$ of terminal sets is used to provide the termination conditions for the deductive process in $Rs$. The successful termination of a refutation process is indicated by the deduction of one of the sets of $\mathcal{F}$: given a finite and clean set $X$ of formulas,

(T1) *$X$ is refutable in $Rs$ iff a set $V \in \mathcal{F}$ can be deduced from $X$ using the inference rules of $Rs$.*

As every set in $\mathcal{F}$ is $C$-inconsistent, the deductive process terminates successfully as soon as an explicit contradiction (represented by any of the sets in $\mathcal{F}$) is deduced from $X$. During the refutation process the right-hand side of the test (T1) should be checked every time a new verifier is being added to the set of deduced formulas. That is, with every deduction of a verifier, we should search $\mathcal{F}$ to see if the verifiers deduced so far form one of such terminal sets. Since the family $\mathcal{F}$, as well as the members of $\mathcal{F}$, can be large (for example, the family of terminal sets of $Rs_3$ consists of 54 elements), the verification of the right-hand side of (T1) may cause a reasoning program to perform a substantial number of additional steps during the deductive process. For some logics, however, the termination condition (T1) can be replaced by a simpler test: for some finite set $V$ of verifiers of $Rs$,

(T2) *$X$ is refutable in $Rs$ iff a verifier from $V$ can be deduced from $X$ using the inference rules of $Rs$.*

In view of (T2), a refutational process terminates successfully as soon as one of the members of $V$ is deduced from $X$. In comparison with (T1), a successful termination of the deductive process is 'witnessed' now by a single verifier (i.e., a member of $V$) rather than a set of verifiers (i.e., a member of $\mathcal{F}$). As a result, we can save a substantial number of deductive steps in the search for a refutation. The following example shows that this is possible for the logic $\mathcal{P}_3$.

EXAMPLE 4.20: Let us consider the Łukasiewicz logic $\mathcal{P}_3$ and its resolution counterpart $Rs_3 = \langle \mathcal{V}_3, \mathcal{F}_3 \rangle$ (see Appendix B). Clearly, for every finite and clean set $Y$ of formulas,

(T1*) *$Y$ is refutable in $Rs_3$ iff a set $V \in \mathcal{F}_3$ can be deduced from $Y$ using the inference rules of $Rs_3$.*

It turns out that for $\mathcal{P}_3$ and $Rs_3$ a simpler termination test can be proposed:

(T2*) *$Y$ is refutable in $Rs_3$ iff a verifier other than $v_2$ and $v_5$ can be deduced from $Y$ using the inference rules of $Rs_3$.*

The correctness of this test will be discussed in Example 4.24. In the light of (T2*), we may consider all the verifiers other than $v_2$ and $v_5$ to be 'witnesses' of inconsistency. As soon as one of such verifiers is deduced, we have sufficient information to successfully terminate the deductive process. This makes not only the storage of the family $\mathcal{F}_3$ unnecessary, but it also significantly speeds up the refutation process, as the successful refutation is witnessed now by a single verifier rather than a set of verifiers.

To see (T2*) in action, let us construct a refutation of

$$X = \{p_1, p_1 \to p_2, p_2 \to p_3, \neg p_3\}$$

in $Rs_3$. First, we apply the resolution rule to the sequence

$$p_2 \to p_3,\ p_2 \to p_3,\ \neg p_3,\ p_2 \to p_3,\ p_2 \to p_3,\ \neg p_3$$

and obtain $(p_2 \to v_0) \vee (p_2 \to v_1) \vee \neg v_2 \vee (p_2 \to v_3) \vee (p_2 \to v_4) \vee \neg v_5$, which reduces to the formula

$$\gamma(p_2) = (p_2 \to v_0) \vee (p_2 \to v_1) \vee (p_2 \to v_3) \vee (p_2 \to v_4) \vee v_3.$$

Next, we apply the resolution rule to the sequence

$$p_1 \to p_2,\ p_1 \to p_2,\ \gamma(p_2),\ p_1 \to p_2,\ p_1 \to p_2,\ \gamma(p_2)$$

and obtain $(p_1 \to v_0) \vee (p_1 \to v_1) \vee \gamma(p_2/v_2) \vee (p_1 \to v_3) \vee (p_1 \to v_4) \vee \gamma(p_2/v_5)$, which reduces to the formula

$$\delta(p_1) = (p_1 \to v_0) \vee (p_1 \to v_1) \vee (p_1 \to v_3) \vee (p_1 \to v_4) \vee v_4.$$

Finally, we resolve

$$p_1,\ p_1,\ \delta(p_1),\ p_1,\ p_1,\ \delta(p_1)$$

the resolvent of which reduces to $v_4$. Let us note that $\{v_4\}$ is not in $\mathcal{F}_3$ and, hence, (T1*) cannot be used at this point to conclude that the refutation process should be terminated. However, we can make such a conclusion using (T2*); the deduction of $v_4$ terminates successfully the refutation process and the inconsistency of $X$ in $\mathcal{P}_3$ has been confirmed.

Curiously enough, the verifier $v_4$ which has been used to terminate the refutation of $X$ is not a contradictory formula, as it is satisfied in $M_3$ by any valuation $h$ such that $h(p_0) = 0$. This means that for a successful termination of the deductive process we don't necessarily need to deduce an explicit contradiction (represented by unsatisfiable verifiers $v_0$ or $v_1$); as far as (T2*) is concerned, the deduction of the satisfiable verifier $v_3$ or $v_4$ is equally sufficient. ∎

Łukasiewicz logics are not the only calculi for which the termination test (T1) can be replaced by a simpler test (T2). In this section we identify other logics for which resolution counterparts that satisfy (T2) can be found.

LEMMA 4.21: *Let $Rs$ be a resolution counterpart of a disjunctive logic $\mathcal{P} = \langle \mathcal{L}, C \rangle$. Let $h$ be a valuation of $\mathcal{L}$ into the matrix $M_{Rs}$ induced by $Rs$ and let $d$ be a set of designated truth-values of $M_{Rs}$. Then, for every finite set $X \subseteq L$ clean in $Rs$, $h(X) \subseteq d$ implies $r^*(X) \subseteq d$.*

*Proof:* Let $Rs, \mathcal{P}, h$, and $d$ be as stated. Let $X \subseteq L$ be finite, clean in $Rs$, and such that $h(X) \subseteq d$. Let us select $v \in r^*(X)$. Without any loss of generality we can assume that $v$ is obtained from a formula $\alpha$ using only the transformation rules of $Rs$, while $\alpha$ is deduced from $X$ using the resolution rule only (cf. Lemma 2.6). By Lemma 3.17, there is $\beta(p_1, \ldots, p_s) \in X$ such that

$$\alpha = \gamma_0 \vee \beta(p_1/h(p_1), \ldots, p_s/h(p_s)) \vee \gamma_1.$$

Let $\gamma_0^*$ and $\gamma_1^*$ be obtained from $\gamma_0$ and $\gamma_1$, respectively, using the transformation rules of $Rs$ (we can assume that $Var(\gamma_0, \gamma_1) \subseteq Var(Ver)$). Clearly, $\beta(p_1/h(p_1), \ldots, p_s/h(p_s))$ reduces to $h(\beta)$. Hence, $v\ \Theta^R_{C_M}\ (\gamma_0^* \vee h(\beta) \vee \gamma_1^*)$. Since $h(\beta) \in d$, by Lemma 3.10, $v \in d$. ∎

THEOREM 4.22: *Let $Rs = \langle \mathcal{V}, \mathcal{F} \rangle$ be a resolution counterpart of a disjunctive logic $\mathcal{P} = \langle \mathcal{L}, C \rangle$ and let $d$ be a maximal consistent set of verifiers of $Rs$ such that the logic defined by $\langle \mathcal{V}, \{d\} \rangle$ has the same inconsistent sets as $\mathcal{P}$. Then, for every finite and clean $X \subseteq L$,*

(i) $C(X) = L$ *iff* $r^*(X) \not\subseteq d$.

Clearly, the termination condition described in (i) is much easier to test than the one described in (T1).

*Proof:* Let $Rs, \mathcal{P}$, and $d$ be as stated, and let $X \subseteq L$ be finite and clean in $Rs$. If $C(X) \neq L$, then for some $h \in Hom(\mathcal{L}, \langle \mathcal{V}, \{d\} \rangle), h(X) \subseteq d$. By Lemma 4.21, $r^*(X) \subseteq d$. Conversely, if $C(X) = L$, then, by (T1), there is $V \in \mathcal{F}$ such that $V \subseteq r^*(X)$. If $r^*(X) \subseteq d$, then we would have $V \subseteq d$, which is impossible by (r3) and Lemma 3.1(i). ∎

COROLLARY 4.23: *Let $M$ be as in Theorem 4.17. Then for every resolution counterpart $Rs$ of $\langle \mathcal{L}, C_M \rangle$ we can select a maximal consistent set $d$ of verifiers of $Rs$ such that for every finite and clean $X \subseteq L$,*

(i) $C_M(X) = L$ *iff* $r^*(X) \not\subseteq d$.

*Proof:* Follows from Lemma 4.19 and Theorem 4.22. ∎

Among the logics that satisfy the assumptions of Corollary 4.23 there is classical propositional logic, Słupecki's logic $\mathcal{P}_s$, as well as all finitely-valued logics of Łukasiewicz and Post.

EXAMPLE 4.24: In this example, we justify the correctness of the termination test (T2*) (formulated in Example 4.20) for $\mathcal{P}_3$ and its resolution counterpart $Rs_3 = \langle \mathcal{V}_3, \mathcal{F}_3 \rangle$.

First, let us note that Corollary 4.23 is applicable to the Łukasiewicz matrix $M_3$. Indeed, $M_3$ satisfies Theorem 4.17(i). By Exercise 2.7, Theorem 4.17(ii) holds as well.

By Corollary 4.23, one of the maximal consistent sets of verifiers of $Rs_3$ can be used to test the termination of the deductive process. By Theorem 4.22, a maximal consistent set $d$ of verifiers of $Rs_3$ satisfies

(a) $C_3(X) = L$ *iff* $r^*(X) \not\subseteq d$,

for every finite and clean $X \subseteq L$, provided that the logic defined by $\langle \mathcal{V}_3, \{d\} \rangle$ has the same inconsistent sets as $\mathcal{P}_3$. There are only two maximal consistent sets of verifiers of $Rs_3$: $d_0 = \{v_3, v_4, v_5\}$ and $d_1 = \{v_2, v_5\}$. We claim that the matrix $\langle \mathcal{V}_3, \{d_1\} \rangle$ is what we need. For, let $h$ be a valuation of $\mathcal{L}$ into $M_3$ such that for every variable $p$, $h(p) = 1$. Define the equivalence relation $\Theta$ on the set of verifiers of $Rs_3$ in the following way: for every $v, w \in Ver$,

$$v\Theta w \text{ iff } h(v) = h(w).$$

It is easy to check that $\Theta$ is a congruence of $\langle \mathcal{V}_3, \{d_1\} \rangle$ and that the quotient matrix $\langle \mathcal{V}_3, \{d_1\} \rangle / \Theta$ and $M_3$ are isomorphic. Now, our claim follows from Propositions 1.8 and 1.9. In sum, we have demonstrated that (a), rewritten as: for every finite and clean $X \subseteq L$,

(T2*) $C_3(X) = L$ *iff a verifier other than* $v_2$ *and* $v_5$ *can be deduced from* $X$ *in* $Rs_3$,

expresses the termination condition for the deductive process in $Rs_3$. ∎

Lemma 4.10, Theorem 4.17, and Corollary 4.23 describe a class of resolution logics and two techniques which can be used to simplify the resolution based deductive process for these logics. The question that remains to be answered is whether these two techniques can be used simultaneously. The affirmative answer is given in our last theorem of this chapter.

THEOREM 4.25: *Let a matrix $M$ be as in Theorem 4.17 and let $n$ be the cardinality of $M$. Moreover, let $Rs$ be a resolution counterpart of $\langle \mathcal{L}, C_M \rangle$. Then for some selection $v_0, \ldots, v_{n-1}$ of verifiers of $Rs$ and some maximal consistent set $d$ of verifiers of $Rs$, the following holds: for every finite and clean set $X \subseteq L$,*

(i) *$C_M(X) = L$ iff a verifier $v \notin d$ can be deduced from $X$ using the transformation rules of $Rs$ and the following simplified form of the resolution rule*

$$Res_n : \frac{\alpha_0(p), \ldots, \alpha_{n-1}(p)}{\alpha_0(p/v_0) \vee \ldots \vee \alpha_{n-1}(p/v_{n-1})}.$$

*Moreover, $d$ and the verifiers $v_0, \ldots, v_{n-1}$ can be selected effectively.*

*Proof:* Let $M = \langle \mathcal{A}, \{d_M\} \rangle$ and $Rs = \langle \mathcal{V}, \mathcal{F} \rangle$ be as stated, and let $d$ be a maximal consistent set of verifiers of $Rs$ such that the logics defined by $M$ and $\langle \mathcal{V}, \{d\} \rangle$ have the same inconsistent sets. The existence of $d$ is guaranteed by Lemmas 4.18 and 4.19. Let $h^* \in Hom(\mathcal{L}, M)$ be such that $h^*(d) \subseteq d_M$. We claim that

(a) $h^*$ maps $Ver$ onto the set of truth-values of $M$.

By Lemma 4.21, for every finite, clean, and $C_M$-consistent set $X \subseteq L$, $h^*(r^*(X)) \subseteq d_M$. Since $M$ satisfies Theorem 4.17(i), (a) follows from Lemma 4.19.

Now, we are ready to prove (i). Let $\{v_0, \ldots, v_{n-1}\}$ be the set of $n$ verifiers that are mapped by $h^*$ onto the set of truth-values of $M$. Let $X \subseteq L$ be finite and clean. By $r_n(X)$ let us denote the set of all verifiers of $Rs$ that can be deduced from $X$ using $Res_n$ and the transformation rules of $Rs$. If $C_M(X) = L$, then, by Lemma 4.10, for some $V \in \mathcal{F}, V \subseteq r_n(X)$. If $r_n(X) \subseteq d$, then we would have $V \subseteq d$, which is impossible in the light of Lemma 3.1(i) and (r3).

Now, let us assume that $C_M(X) \neq L$. Having (a), we repeat the second part of the proof of Lemma 4.10 to conclude that $h^*(r_n(X)) \subseteq d_M$. This gives us $r_n(X) \subseteq d$ and the proof of (i) has been concluded.

The fact that $d$ and $v_0, \ldots, v_{n-1}$ can be selected effectively is left to the reader (as Exercise 4.8). ∎

EXAMPLE 4.26: Let us consider the resolution counterpart $Rs_5 = \langle \mathcal{V}_5, \mathcal{F}_5 \rangle$ of the logic $\mathcal{P}_5$ (see Appendix D). As in Example 4.24, we show that the termination test (T1) can be replaced by the test: for every finite and clean $X \subseteq L$,

(T2*) $C_5(X) = L$ *iff a verifier* $v$ *other than* $v_4$ *and* $v_9$ *can be deduced from* $X$ *in* $Rs_5$.

Let $h^* \in Hom(\mathcal{L}, M_5)$ be such that $h(p_0) = 1$. Then $h^*(\{v_4, v_9\}) = \{4\}$ and $h^*(\{v_0, \ldots, v_4\}) = \{0, \ldots, 4\}$. Hence, looking at the proof of Theorem 4.25, we can replace the resolution rule of $Rs_5$ by

$$Res_5 : \frac{\alpha_0(p), \ldots, \alpha_4(p)}{\alpha_0(p/v_0) \vee \ldots \vee \alpha_4(p/v_4)}.$$

Let us return to the refutation of $X = \{p_1, p_1 \to p_2, p_2 \to p_3, \neg p_3\}$ in $Rs_5$ presented in Example 4.16. In that example, the inconsistency of $X$ was determined by deducing two verifiers $v_8$ and $v_4$ from $X$. By Theorem 4.25, the deduction of $v_4$ is superfluous and we can successfully terminate the refutation process after the deduction of $v_8$. ∎

**EXERCISES**

4.1. Let $Rs_i = \langle \mathcal{V}_i, \mathcal{F}_i \rangle, i = 0, 1$, be two minimal resolution counterparts of a resolution logic $\mathcal{P}$. Are the algebras $\mathcal{V}_0$ and $\mathcal{V}_1$ necessarily isomorphic?

4.2. Let $\mathcal{P} = \langle \mathcal{L}, C \rangle$ be a resolution logic of $v$-degree $k$. Show that for every $X \subseteq L, C(X) = L$ *iff* $C(e(X)) = L$, for every substitution $e$ that maps $L$ into $L^{(k)}$.

4.3. Show that there is an effective method of verifying whether or not two proper resolution proof systems on $\mathcal{L}$ are resolution counterparts of the same resolution class.

4.4. Let $M$ be a finite and proper matrix for $\mathcal{L}$ of cardinality $m$. Moreover, let $k$ be the cardinality of $L^{(m-1)}/\Theta_M$, where $\Theta_M$ is defined as in Section 4.4. Show that every minimal resolution counterpart of $\langle \mathcal{L}, C_M \rangle$ has at most $k$ verifiers and that $k$ can be effectively computed.

4.5. (E. Harley) Show that the assumption that $M$ has only one designated set of truth-values can be dropped from Lemma 4.10.

4.6. Prove the correctness of the following algorithm that determines whether or not a submatrix $N$ of a finite and proper matrix $M$ for $\mathcal{L}$ defines the logic that has the same inconsistent sets as the logic defined by $M$ (cf. Theorem 4.17).

*Step 1:* Let $k = card(M)$. Construct a sequence $\alpha_0, \ldots, \alpha_s$ of formulas of $L^{(k)}$ such that for every $\beta \in L^{(k)}$ there is $i \leq s$ such that $\beta \Theta_M \alpha_i$.

*Step 2:* Construct the family $\mathcal{D}$ of all maximal $C_M$-consistent subsets of $\{\alpha_0, \ldots, \alpha_s\}$.

*Step 3:* If every $d \in \mathcal{D}$ is $C_N$-consistent, then the logics defined by $M$ and $N$ have the same inconsistent sets. Otherwise, $M$ and $N$ do not have the property in question.

4.7. Let $M$ be as in Theorem 4.17 and let $Rs$ be a resolution counterpart of $\langle \mathcal{L}, C_M \rangle$. By Lemma 4.10, the resolution rule $R^{\vee}_{Ver}$ of $Rs$ can be replaced by the rule $Res_n$ described in this lemma. Let $X \subseteq L$ be finite, clean in $Rs$, and $C_M$-inconsistent. Show that if $T$ is a refutation of $X$ constructed using $R^{\vee}_{Ver}$, then there is a refutation $T_n$ of $X$ constructed using $Res_n$, and, in this refutation, $Res_n$ has been applied the same number of times as $R^{\vee}_{Ver}$ in $T$. Prove that the converse of this fact is also true: if $T_n$ is a refutation of $X$ constructed using $Res_n$, then there is a refutation $T$ of $X$ constructed using $R^{\vee}_{Ver}$ in which there are as many applications of $R^{\vee}_{Ver}$ as there are applications of $Res_n$ in $T_n$.

4.8. An implementation of a theorem proving procedure based on Theorem 4.25 requires an effective selection of a maximal consistent set $d$ of verifiers and of a list $v_0, \ldots, v_{n-1}$ of verifiers described in this theorem. Design an algorithm that selects $d$ and $v_0, \ldots, v_{n-1}$.

# 5

# THEOREM PROVING STRATEGIES

In Chapter 4 we discussed ways of improving the efficiency of resolution based deductive process by minimizing the length of resolvents and by simplifying the termination test. In this chapter we focus on theorem proving strategies – the speed-up techniques for controlling and directing the deductive process of an automated reasoning system.

The search for powerful and efficient strategies for directing the reasoning program in its choice of formulas and inference rules at any step of the deductive process as well as for the discarding of redundant information accumulated during such a process is recognized as one of the central issues in the area of Automated Reasoning (cf. [77]).

> Even for the simplest of problems, an uncontrolled use of some chosen inference rule will produce far too much information. ([77]).

An efficient resolution based reasoning program must incorporate some theorem proving strategies to control the application of its main inference rule – the resolution rule. The generation of resolvents that would evidently have no impact on the completion of a reasoning task must be blocked.

Some of the speed-up techniques which were developed for classical logic can be generalized and applied to resolution logics. In this chapter we study two of such techniques, the set of support strategy and the polarity strategy, and investigate their impact on the reduction of the size of the search space. Other well-known strategies, such as *weighting strategy* (cf. [33]) or *subsumption* (cf. [54]), can also be adopted to our resolution framework without substantial effort.

For reasons of simplicity, we assume that all resolution logics discussed in this chapter are disjunctive.

## 5.1 SET OF SUPPORT STRATEGY

The *set of support strategy* was proposed by Wos, Robinson, and Carson in [78]. Informally speaking, this strategy prohibits the resolution of formulas that belong to a certain consistent set. In this section we generalize this strategy to resolution logics.

Let $Rs = \langle \mathcal{V}, \mathcal{F} \rangle$ be a resolution counterpart of a disjunctive logic $\mathcal{P} = \langle \mathcal{L}, C \rangle$. A subset $Y$ of a set $X$ of formulas is called a *set of support* of $X$ if $X - Y$ is $C$-consistent. An instance

$$\frac{\alpha_0(p), \ldots, \alpha_n(p)}{\alpha_0(p/v_0) \vee \ldots \vee \alpha_n(p/v_n)}$$

of the resolution rule of $Rs$ is said to be $(X, Y)$-supported if $Y$ is a set of support of $X$ and $\alpha_i(p) \notin X - Y$, for some $i \leq n$. We say that a refutation in $Rs$ is $(X, Y)$-supported, if every application of the resolution rule is $(X, Y)$-supported. In other words, the set of support strategy prohibits the resolution rule from being applied to a sequence of formulas when all these formulas are from the consistent set $X - Y$.

The following theorem shows that $Rs$ retains its refutational completeness with respect to $\mathcal{P}$ when we require every refutation to be $(X, Y)$-supported.

THEOREM 5.1 (O'Hearn, Stachniak): *The set of support strategy is complete. That is, if $Y$ is a set of support of a finite $C$-inconsistent set $X$ clean in $Rs$, then there is an $(X, Y)$-supported refutation of $X$ in $Rs$.*

*Proof:* Let $X$ and $Y$ be as stated and let $M_{Rs} = \langle \mathcal{V}, \mathcal{D} \rangle$ be the matrix induced by $Rs$. Since $X$ is $C$-inconsistent and since $C_{Rs}$ has the same inconsistent sets as $C$, $C_{Rs}(X) = L$. Now, we repeat the argument presented in the proof of Lemma 3.13 with one modification: when turning an inference node into a failure node we must ensure that the application of the resolution rule satisfies the set of support restriction.

Suppose that the refutation (tree) $T$ has been constructed until level $k \geq 1$. Let $S_0^N, \ldots, S_m^N$ be the closed semantic trees associated with the (only) node $N$ of $T$ at level $k$. If all these trees are one-element, then we stop the construction of $T$. Otherwise, there is a tree $S_j^N$ and an inference node $K$ such that

(a) some child $K_j$ of $K$ falsifies a formula $\alpha_j(p_{k+1}) \notin X - Y$ in $\langle \mathcal{V}, \{d_j\} \rangle$.

To justify (a), let us assume, on the contrary, that $S_j^N$ and $K$ satisfying (a) cannot be found. Let $S_j^N$ be any tree with an inference node $K$. Since every child of $K$ falsifies a formula from $X - Y$, $S_j^N$ is a closed semantic tree of $X - Y$ with respect to $\langle \mathcal{V}, \{d_j\} \rangle$. So, for every $d_j \in \mathcal{D}$, $X - Y$ is $C_{\langle \mathcal{V}, \{d_j\} \rangle}$-inconsistent. This means, however, that $X - Y$ is $C_{Rs}$-inconsistent contradicting our initial assumption concerning $X - Y$. Hence, a tree $S_j^N$ and an inference node $K$ of $S_j^N$ satisfying (a) exist. The rest of the proof can be copied from the proof of Lemma 3.13. ∎

EXAMPLE 5.2: Let us return to the refutation of the set $X = \{p_1, p_1 \to p_2, p_2 \to p_3, \neg p_3\}$ in $Rs_3$ presented in Example 4.20. By inspecting the deductive steps we conclude that this refutation is $(X, \{\neg p_3\})$-supported. In this case, the set of support strategy blocks 129 out of 192 initial applications of the resolution rule, leaving us with 63 possible choices. To block further applications of the resolution rule that do not contribute significantly to the search for a refutation, we need to combine the set of support with other speed-up techniques such as the polarity strategy discussed in the following sections of this chapter. ∎

An important practical issue is the selection of a set of support for a particular reasoning task. By (ref), the validity of an inference $\alpha \in C(X)$ is determined by checking the consistency status of the deduced set $X_\alpha$. Frequently we know that $X$ is $C$-consistent. If, in addition, $X_\alpha = X \cup Y$, then, evidently, $Y$ is one of the possible choices for a set of support. For instance, for the $n$-valued Łukasiewicz logic $\mathcal{P}_n$, $X_\alpha = X \cup \{\sim \alpha\}$ and we may always select $\{\sim \alpha\}$ as a set of support (cf. Example 1.6). In general, however, $X$ may not be a subset of $X_\alpha$ and some other criteria for selecting a suitable set of support have to be used. For more discussion concerning the set of support strategy the reader is referred to [78, 77].

## 5.2 POLARITY STRATEGY

The second speed-up technique discussed in this chapter is the polarity strategy. The central idea behind this restriction technique is to predict the 'usefulness' of a given resolvent, basing our assessment on the arrangement of formulas which we want to resolve and on the choice of a variable to be resolved upon. The formation of 'useless' resolvents should be prohibited.

In some cases it is easy to make such predictions. For instance, we should certainly block the resolution of any sequence of six formulas $\alpha_0(p), \ldots, \alpha_5(p)$ in $Rs_3$

when $\alpha_5$ is a variable, i.e., when $\alpha_5 = p$. Indeed, the resolvent

$$\alpha_0(p/v_0) \vee \ldots \vee \alpha_4(p/v_4) \vee v_5,$$

being a tautology of $\mathcal{P}_3$, is useless from the standpoint of the deductive process. The polarity strategy will help us to block the formation of resolvents in many other non-trivial cases.

To formulate the polarity strategy we require a partial order $\leq$ of the set of all formulas of a given language so that $\alpha \leq \beta$ only if $\beta$ is, in a certain sense, 'logically larger' than $\alpha$. For example, '$\alpha \leq \beta$' may mean that, in a given logic $\langle \mathcal{L}, C \rangle$, $\beta \in C(\alpha)$. We call $\leq$ a *polarity relation*. Then, we effectively identify some subformulas of formulas as positive ('+') or negative ('−') in such a way that if a subformula $\alpha$ of $\beta$ is positive in $\beta$, then the replacement of $\alpha$ with a larger expression (with respect to $\leq$) results in a larger formula. Similarly, if $\alpha$ is negative in $\beta$, then any replacement of $\alpha$ with a smaller expression (with respect to $\leq$) results in a larger formula. Finally, equipped with an algorithm for identifying positive and negative subformulas, we can speed-up the deductive process by selecting and arranging formulas for resolution in such a way that positive occurrences of the variable (or subformula) being resolved upon are replaced by 'small' verifiers (with respect to $\leq$) while negative occurrences are replaced by 'large' verifiers. As a result, many useless applications of the resolution rule will be blocked.

The restrictive power of the polarity strategy depends on the choice of a polarity relation as well as on a polarity assignment algorithm (some subformulas may be positive, negative, or both with respect to one polarity relation or one polarity assignment method, whereas they can be of no polarity with respect to other polarity relations and polarity assignment methods).

A good choice for a polarity relation for classical propositional logic $\mathcal{P}_2$ is the relation defined by the equivalence

(pol) $$\alpha \leq \beta \textit{ iff } \beta \in C_2(\alpha).$$

A simple recursive algorithm can be used to assign polarity values to subformulas of formulas of this logic in the way described in [30, 39]. The polarity strategy used by Murray in his non-clausal resolution proof system for $\mathcal{P}_2$ blocks applications of the resolution rule to all pairs of formulas that do not satisfy certain constraints formulated in terms of the polarity values of the variable being resolved upon (cf. [39]). For instance, this strategy prohibits any pair of formulas from being resolved upon a variable which is only positive or only negative in both formulas.

Every resolution proof system $Rs$ on $\mathcal{L}$ defines a polarity relation $\leq_{Rs}$ by, roughly speaking, replacing the consequence operation $C_2$ in (pol) by the consequence operation $C_{Rs}$ defined by the matrix induced by $Rs$. The formal definition of the polarity relation $\leq_{Rs}$ requires the introduction of the notion of an interpretation substitution.

Let $Rs$ be a resolution counterpart of a disjunctive logic $\mathcal{P}$ and let $M_{Rs}$ be the matrix induced by $Rs$. We say that a logical substitution $e$ is an *interpretation substitution* if for every variable $p$ occurring in the verifiers of $Rs, e(p) = p$, while for the remaining variables, $e(p)$ is a verifier of $Rs$. Clearly, if $e$ is an interpretation substitution, then for every formula $\alpha, e(\alpha)$ can be reduced to an irreducible formula $e^*(\alpha)$ using the transformation rules of $Rs$. By Corollary 2.10, we can assume that for every interpretation substitution $e$, $e^*$ is an identity function on the set of verifiers of $Rs$, i.e., for every verifier $v, e^*(v) = v$. In other words, we treat verifiers as logical constants with respect to interpretation substitutions. Now, we are ready to define the polarity relation $\leq_{Rs}$. For every two formulas $\alpha, \beta \in L$,

> $\alpha \leq_{Rs} \beta$ *iff* for every set $d$ of designated truth-values of $M_{Rs}$ and every interpretation substitution $e$, if $e^*(\alpha) \in d$, then $e^*(\beta) \in d$.

In other words, if $\alpha \leq_{Rs} \beta$, then every interpretation substitution $e^*$ that 'designates' $\alpha$ in $M_{Rs}$ 'designates' $\beta$ as well, or, informally stated, $\beta$ is as true in $M_{Rs}$ as $\alpha$. We shall write $\alpha =[Rs]\ \beta$ if $\alpha \leq_{Rs} \beta$ and $\beta \leq_{Rs} \alpha$.

EXAMPLE 5.3: Let us consider the resolution counterpart $Rs_3$ of the 3-valued Łukasiewicz logic ($Rs_3$ is defined in Section 4.3, see also Appendix B). Consider the formulas:

$$\alpha = v_0 \vee p_1,$$
$$\beta = v_5 \vee p_1.$$

We can easily verify that $\alpha \leq_{Rs_3} \beta$. Indeed, for every interpretation substitution $e, e^*(v_5 \vee p_1) = v_5$ and $v_5$ belongs to every maximal consistent set of verifiers of $Rs_3$. On the other hand, $\beta \leq_{Rs_3} \alpha$ does not hold. Clearly, if $e$ is the interpretation substitution such that $e(p_i) = v_0$, all $i > 0$, and if $d = \{v_2, v_5\}$, then $e^*(\beta) = e^*(v_5 \vee v_0) = v_5 \in d$ while $e^*(\alpha) = e^*(v_0 \vee v_0) = v_0 \notin d$. Now, our claim follows at once from the fact that $d$ is maximal consistent in $Rs$. ∎

Once the polarity relation has been chosen, the next step is to provide an algorithm $\mathcal{H}$ for assigning polarity values to subformulas of formulas. When this is done, the polarity strategy for $Rs$ with respect to $\mathcal{H}$ can be defined as follows.

(pol1) *The resolvent $\beta_0(p/v_0) \vee \ldots \vee \beta_n(p/v_n)$ of $\beta_0, \ldots, \beta_n$ cannot be formed and used during the deductive process if there are $i, j \leq n$ such that $v_i \leq_{Rs} v_j$ and either*

- *$p$ is positive in $\beta_j$ but not in $\beta_i$*

*or*

- *$p$ is negative in $\beta_i$ but not in $\beta_j$.*

We say that $Rs$ admits polarity strategy (pol1), if this system remains refutationally complete for $\mathcal{P}$ while its resolution rule is restricted by (pol1).

EXAMPLE 5.4. In this example we discuss (pol1) using the 3-valued Łukasiewicz logic $\mathcal{P}_3$ and its resolution counterpart $Rs_3$ (cf. Appendix B). The polarity relation $\leq_{Rs_3}$, restricted to the set $\{v_0, \ldots, v_5\}$ of verifiers of $Rs_3$, is graphed in Figure 5.1. For every two verifiers $v_i \neq v_j, v_i \leq_{Rs_3} v_j$ if and only if there is a directed path from $v_i$ to $v_j$.

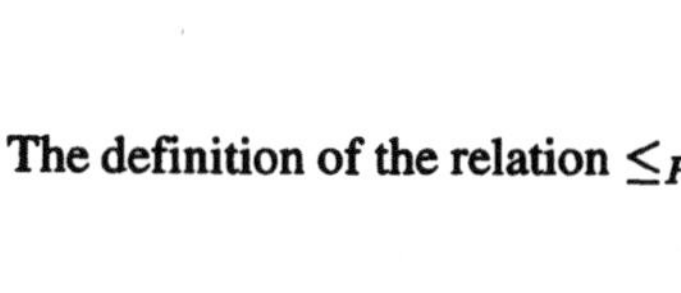


FIG. 5.1: The definition of the relation $\leq_{Rs_3}$ for $Rs_3$.

Let

(a) $$\beta_0(p),\ \beta_1(p),\ \beta_2(p),\ \beta_3(p),\ \beta_4(p),\ \beta_5(p)$$

be a sequence of formulas which we want to resolve upon $p$ using the resolution rule of $Rs_3$ restricted by (pol1). Depending on a polarity assignment algorithm, the variable $p$ may assume any of the polarity values, or be of no polarity at all, in any of these formulas. For the reasons of simplicity, let us assume that for no $j \leq 5$, $p$ is both positive ('+') and negative ('−') in $\beta_j$. The polarity strategy (pol1) allows the formation of the resolvent

$$\beta_0(p/v_0) \vee \beta_1(p/v_1) \vee \beta_2(p/v_2) \vee \beta_3(p/v_3) \vee \beta_4(p/v_4) \vee \beta_5(p/v_5),$$

provided that the polarity values of $p$ in the formulas of (a) form one of the 'admissible' patterns. To explain what these admissible patterns are, let

$$\text{(b)} \qquad i_0 \ i_1 \ i_2 \ i_3 \ i_4 \ i_5$$

be the sequence that records the polarity values of $p$ in the formulas of (a), i.e., for every $j \leq 5$, $i_j$ is the polarity value of $p$ in $\beta_j$. It can be easily verified that (pol1) admits, for instance, the pattern

$$+ \ + \ + \ + \ + \ +,$$

i.e., (pol1) will not block the resolution of (a) when $p$ is only positive in every $\beta_i$. On the other hand, (pol1) will block the resolution of (a), if (b) is

$$- \ + \ + \ + \ + \ +.$$

Indeed, $v_0 \leq_{Rs_3} v_1$. Since $p$ is negative in $\beta_0$ and positive in $\beta_1$, (pol1) is violated.

There are twenty patterns of polarity values of $p$ which are consistent with (pol1). If, in addition, we assume that $p$ is only '+' or only '−' in every formula of (a), then we are left with six patterns:

$$\begin{array}{cccccc}
- & - & - & - & - & - \\
+ & + & - & - & - & - \\
+ & + & - & + & + & - \\
+ & + & + & - & - & - \\
+ & + & + & + & + & - \\
+ & + & + & + & + & +.
\end{array}$$

Under this assumption, we can apply the resolution rule of $Rs_3$ to sequence (a) without violating (pol1) only if (b) is among the six patterns listed above. ■

In the subsequent sections we discuss three polarity assignment algorithms and investigate their effect on the reduction of the size of the search space.

## 5.3 OPERATOR POLARITY

We begin our discussion on polarity assignment algorithms with a variant of the polarity assignment method used by Murray in [39] for his non-clausal resolution proof system for classical logic and further developed by Manna and Waldinger in [30]. At the core of this method is the assignment of polarity values to argument occurrences of logical connectives.

Let $\mathcal{P}$, $Rs$, and $M_{Rs}$ be as in the preceding section. By $L^*$ let us denote the set of all formulas $\alpha$ of $\mathcal{L}$ that may occur in a refutation of a clean set. In general, $L^* \neq L$. For example, the formula $p_0 \vee p_1$ will never occur in a refutation in the resolution counterpart $Rs_2$ of classical propositional logic. Indeed, since $p_0$ occurs in the verifiers of $Rs_2$, this formula can only be resolved upon $p_1$, and neither $p_0 \vee v_0$ nor $p_0 \vee v_1$ transforms into a verifier of $Rs_2$. So, $p_0 \vee p_1 \notin L^*$. $L^*$ can be formally defined as the set of all formulas $\alpha$ of $L$ such that for every interpretation substitution $e, e^*(\alpha)$ is a verifier of $Rs$. In this chapter, we choose to work with $L^*$ rather than with the whole set $L$, since during the deductive process in $Rs$ only formulas from $L^*$ are used. Moreover, the restriction to $L^*$ will allow us to simplify many arguments during our discussion on the polarity strategy.

Let $f$ be a $k$-ary connective of $\mathcal{L}$. We say that $f$ is *positive* over its $i$-th argument, $1 \leq i \leq k$, if and only if for every $\alpha, \alpha_1, \ldots, \alpha_k \in L^*$,

$$\text{(op+)} \quad \alpha_i \leq_{Rs} \alpha \text{ implies } f(\alpha_1, \ldots, \alpha_k) \leq_{Rs} f(\alpha_1, \ldots, \alpha_{i-1}, \alpha, \alpha_{i+1}, \ldots, \alpha_k).$$

Similarly, we say that $f$ is *negative* over its $i$-th argument if and only if for every $\alpha, \alpha_1, \ldots, \alpha_k \in L^*$,

$$\text{(op−)} \quad \alpha \leq_{Rs} \alpha_i \text{ implies } f(\alpha_1, \ldots, \alpha_k) \leq_{Rs} f(\alpha_1, \ldots, \alpha_{i-1}, \alpha, \alpha_{i+1}, \ldots, \alpha_k).$$

We say that $f$ is of no polarity over its $i$-th argument if it is neither positive nor negative. Following [30], we summarize the definition of the notion of a positive/negative occurrence of a subformula as follows. Let $\alpha$ be a subformula of $\beta$.

(p1) If $\alpha = \beta$, then $\alpha$ occurs positively in $\beta$ (i.e., every formula is positive in itself).

(p2) Suppose $\beta = f(\beta_1, \ldots, \beta_k)$ and let us consider a fixed occurrence of $\alpha$ in some $\beta_i, 1 \leq i \leq k$. Then:

- if this occurrence has no polarity value in $\beta_i$ or $f$ has no polarity over its $i$-th argument, then this occurrence has *no polarity* in $\beta$;
- this occurrence is *positive* in $\beta$ if either it is positive in $\beta_i$ and $f$ is positive over its $i$-th argument, or it is negative in $\beta_i$ and $f$ is negative over its $i$-th argument;
- in the remaining cases, this occurrence is *negative* in $\beta$.

EXAMPLE 5.5: Let us consider classical propositional logic $\mathcal{P}_2$ expanded by the addition of the equivalence connective $\equiv$. The syntactic structure of every formula of this logic can be conveniently represented as a finite tree in which every leaf is labelled with a propositional variable while every internal node is labelled with a logical connective. For example, the formula $\beta = (\neg p_0 \vee (p_1 \wedge p_2)) \rightarrow (p_1 \equiv p_0)$ is represented by the tree $T_\beta$ graphed in Figure 5.2. In this representation, an occurrence of a subformula of $\beta$ is just a subtree of $T_\beta$ and, hence, we can talk about assigning polarity values to nodes of $T_\beta$ (i.e., to the roots of such subtrees).

The matrix induced by the two-verifier resolution counterpart $Rs_2$ of $\mathcal{P}_2$ is just the two-element matrix defining this logic. (Obviously, we have to expand the set of transformation rules of $Rs_2$ by the addition of the transformation rules for $\equiv$.) Thus, the polarity relation $\leq_{Rs_2}$ can be equivalently defined as in (pol), i.e., for every two formulas $\alpha$ and $\beta$,

$$\alpha \leq_{Rs_2} \beta \textit{ iff } \beta \text{ is deducible from } \alpha \text{ in } \mathcal{P}_2.$$

It can be easily verified that:

$\neg$ is negative over its argument;

$\vee$ and $\wedge$ are positive over both their arguments;

$\rightarrow$ is negative over its first and positive over its second argument;

$\equiv$ is of no polarity over any of its arguments.

The tree in Figure 5.2 contains all the relevant information concerning the polarities of the occurrences of subformulas of $\beta$. Every edge that joins a non-leaf node with one of its children is labelled with '+', '−', or '0' (no polarity) in accordance with (op+) and (op−).

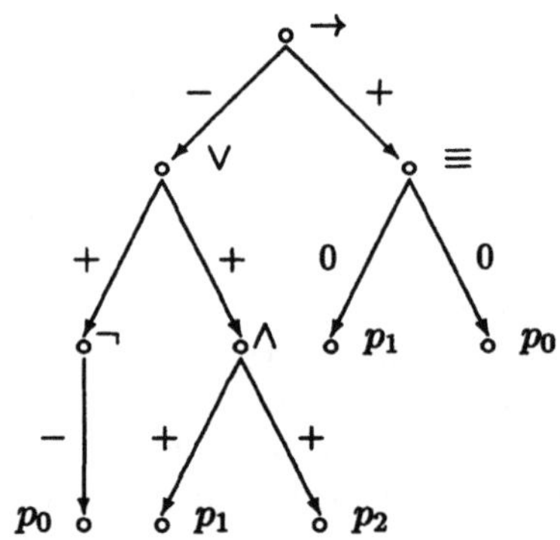


FIG. 5.2: Tree representation of $\beta = (\neg p_0 \vee (p_1 \wedge p_2)) \rightarrow (p_1 \equiv p_0)$.

From (p2) it follows that an occurrence of a subformula of $\beta$, represented by the root $N$ of an appropriate subtree of $T_\beta$, is positive in $\beta$, if the path from the root of $T_\beta$ to $N$ contains no edges labelled with 0 and an even number of edges labelled with '−'. If the number of such edges is odd, then this occurrence is negative. Hence, $\neg p_0$ and $p_1 \wedge p_2$ occur negatively in $\beta$ while the occurrence of $p_1 \equiv p_0$ is positive. Similarly, the leftmost occurrence of $p_0$ is positive while the rightmost occurrence of this variable is of no polarity as the path from the root of $T_\beta$ to this node has an edge labelled with 0. ∎

THEOREM 5.6: *There is an effective procedure which determines the polarity values of $f$ over its arguments, for every connective $f$.*

*Proof:* Let $f$ be a $k$-ary connective of $\mathcal{L}$. To compute the polarity values of $f$ over its arguments we can use the conditions (op+) and (op−) as tests with the term '$L^*$' replaced by '$Ver_{Rs}$'. To prove this claim, let us consider (op+) and assume that for some $\alpha, \alpha_1, \ldots, \alpha_k \in L^*$,

$\alpha_i \leq_{Rs} \alpha$ but $f(\alpha_1, \ldots, \alpha_k) \leq_{Rs} f(\alpha_1, \ldots, \alpha_{i-1}, \alpha, \alpha_{i+1}, \ldots, \alpha_k)$
does not hold.

This means that for some interpretation substitution $e$ and some set $d$ of designated truth-values of $M_{Rs}$,

(a) $e^*(f(\alpha_1, \ldots, \alpha_k)) \in d$ while $e^*(f(\alpha_1, \ldots, \alpha_{i-1}, \alpha, \alpha_{i+1}, \ldots, \alpha_k)) \notin d$.

Clearly,

(b) $e^*(\alpha_i) \leq_{Rs} e^*(\alpha)$.

Since $e^*(f(e^*(\alpha_1), \ldots, e^*(\alpha_{i-1}), e^*(\alpha), e^*(\alpha_{i+1}), \ldots, e^*(\alpha_k)))$ is the same as $e^*(f(\alpha_1, \ldots, \alpha_{i-1}, \alpha, \alpha_{i+1}, \ldots, \alpha_k))$ and since $e^*(f(e^*(\alpha_1), \ldots, e^*(\alpha_k))) = e^*(f(\alpha_1, \ldots, \alpha_k))$, we also have

(c) $e^*(f(e^*(\alpha_1), \ldots, e^*(\alpha_{i-1}), e^*(\alpha), e^*(\alpha_{i+1}), \ldots, e^*(\alpha_k))) \notin d$ while
$e^*(f(e^*(\alpha_1), \ldots, e^*(\alpha_k))) \in d$.

Since $e^*(\alpha), e^*(\alpha_1), \ldots, e^*(\alpha_k)$ are verifiers of $Rs$, (b) and (c) prove the claim for (op+). In the case of (op−) the proof is similar. ∎

We say that a variable $p$ is *positive* (*negative*) in $\alpha$ if every occurrence of $p$ in $\alpha$ is positive (is negative). A variable is said to be of *no polarity* in $\alpha$ if it is neither positive nor negative in this formula. Henceforth, we shall call this version of polarity the *operator polarity*. When the polarities of arguments of every connective are known, the operator polarity values of all the variables in a formula $\alpha$ can be determined in no more than $n\lfloor log_t\, n\rfloor$ time, where $t$ is the maximal arity among the connectives of $\mathcal{L}$ and $n$ is the length of $\alpha$.

THEOREM 5.7 (O'Hearn, Stachniak): *Let $\alpha(p), \beta_0, \beta_1 \in L^*$ and let $\beta_0 \leq_{Rs} \beta_1$. Then:*

(i) *if $p$ is positive in $\alpha(p)$, then $\alpha(p/\beta_0) \leq_{Rs} \alpha(p/\beta_1)$;*

(ii) *if $p$ is negative in $\alpha(p)$, then $\alpha(p/\beta_1) \leq_{Rs} \alpha(p/\beta_0)$.*

*Proof:* (i) and (ii) can be easily demonstrated by induction on the complexity of $\alpha$. ∎

THEOREM 5.8 (O'Hearn, Stachniak): *$Rs$ admits* (pol1) *with respect to the operator polarity.*

*Proof:* Let $X \subseteq L$ be finite and clean in $Rs$. Clearly, as $Rs$ is a resolution counterpart of $\mathcal{P}$, $C(X) \neq L$ implies that $X$ cannot be refuted in $Rs$.

Let us assume that $C(X) = L$ and let $C_{Rs}$ be the consequence operation defined by the matrix $M_{Rs} = \langle \mathcal{V}, \mathcal{D} \rangle$ induced by $Rs$. So, $C_{Rs}(X) = L$. Now, we repeat the argument presented in the proof of Lemma 3.13 with one modification: when turning an inference node into a failure node we must ensure that the application of the resolution rule does not violate (pol1).

Suppose that the refutation (tree) $T$ has been constructed until level $k \geq 1$. Let $S_0^N, \ldots, S_m^N$ be the closed semantic trees associated with the (only) node $N$ at level $k$. If all these trees are one-element, then we stop the construction of $T$. Otherwise, there is a tree $S_j^N$ with an inference node $K$. Let $\alpha_0(p_{k+1}), \ldots, \alpha_n(p_{k+1})$ be the formulas falsified by the children of $K$. If for some $i, j \leq n$, $v_i \leq_{Rs} v_j$ and if $p_{k+1}$ is positive in $\alpha_j$ but not in $\alpha_i$, then we can replace $\alpha_i$ with $\alpha_j$. Clearly, if $\alpha_j(p_{k+1}/p_{v_i})$ were satisfied at $K_i$, then, by Theorem 5.7, $K_j$ could not falsify $\alpha_j(p_{k+1}/p_{v_j})$. Similarly, if $p_{k+1}$ is negative in $\alpha_i$ but not in $\alpha_j$, then we can replace $\alpha_j$ with $\alpha_i$.

What we need now is to demonstrate that these replacements can be performed in such a way that (pol1) is not violated. To this end, let us select any $\alpha_j$ such that

(a) $p_{k+1}$ is positive in $\alpha_j$ and for some $i \leq n$, $v_i \leq_{Rs} v_j$ but $p_{k+1}$ is not positive in $\alpha_i$.

Then, for every $i \leq n$ such that $v_i \leq_{Rs} v_j$ and $p_{k+1}$ is not positive in $\alpha_i$, let us replace $\alpha_i$ with $\alpha_j$. Repeat this process until no $\alpha_j$ satisfying (a) is left. Next, let us select any $\alpha_i$ such that

(b) $p_{k+1}$ is negative in $\alpha_i$ and for some $j \leq n, v_i \leq_{Rs} v_j$ but $p_{k+1}$ is not negative in $\alpha_j$.

Then, for every $j \leq n$ such that $v_i \leq_{Rs} v_j$ and $p_{k+1}$ is not negative in $\alpha_j$, replace $\alpha_j$ with $\alpha_i$. Repeat this process until no $\alpha_i$ satisfying (b) is left. The proof that the resulting string satisfies (pol1) is left to the reader.

The rest of the proof can be copied from the proof of Lemma 3.13. ∎

EXAMPLE 5.9: In this and the future examples presented in this chapter we consider the resolution proof system $Rs_3$ of the 3-valued Łukasiewicz logic $\mathcal{P}_3$. Let us analyze the effect of (pol1) formulated for the operator polarity on the test sets:

$$X_0 = \{p_3, p_3 \to p_1, p_1 \to p_2, \neg p_2\},$$
$$X_1 = \{p_3, p_3 \to p_1, p_1 \to p_2, \sim p_2\},$$

where '$\sim p_2$' abbreviates '$\neg p_2 \vee ((p_2 \to \neg p_2) \wedge (\neg p_2 \to p_2))$'.

First, let us note that $\neg$ and $\to$ have no (operator) polarity values over their arguments. Hence, by (p2), $p_2$ has no polarity value in $\sim p_2$. However, since $p_3$ is positive in itself and since $v_0 =[Rs_3]\ v_1$ and $v_3 =[Rs_3]\ v_4$, (pol1) blocks, for example, the resolution of any sequence $\alpha_0(p_3), \ldots, \alpha_5(p_3)$ of formulas of these sets such that $\alpha_0 \neq \alpha_1$ or $\alpha_3 \neq \alpha_4$. However, (pol1) does not prevent us from resolving formulas of these sets upon $p_2$. Out of 192 initial applications of the resolution rule to any of these sets, (pol1) blocks 58 initial applications leaving us with 134 (cf. Table 5.1 in Section 5.9). ∎

The proof of Theorem 5.8 shows one way in which the polarity strategy (pol1) can be strengthened. The process of replacement described in that proof not only guarantees that the polarity values of the variable being resolved upon form an 'admissible' pattern but also indicates that some of the formulas which we are resolving should be identical. For example, suppose that we want to resolve

$$\beta_0(p),\ \beta_1(p),\ \beta_2(p),\ \beta_3(p),\ \beta_4(p),\ \beta_5(p)$$

upon $p$ in $Rs_3$. If $p$ is positive in $\beta_5$, then we can require that $\beta_0 = \ldots = \beta_5$, since for every $i \leq 5, v_i \leq_{Rs_3} v_5$. If, however, $p$ is not positive in $\beta_5$ but it is positive in $\beta_4$ and, in addition, if this variable is only negative in $\beta_2$, then the only pattern of polarity values of $p$ which satisfies these assumptions without violating (pol1) is

$$+ \quad + \quad - \quad + \quad + \quad -.$$

By repeating the replacement steps described in the proof of Theorem 5.8, we conclude that, for this pattern of polarity values, (pol1) can be strengthened by demanding that $\beta_0 = \beta_1 = \beta_3 = \beta_4$ and $\beta_2 = \beta_5$.

## 5.4 UNRESTRICTED POLARITY

Although (pol1) formulated for the operator polarity can substantially reduce the search space, for many logics the notion of the operator polarity is not subtle enough to take full advantage of (pol1). It turns out that (pol1) can have a more profound effect on the reduction of the size of the search space if the operator polarity is replaced with the following notion. Let $Rs$ be a resolution counterpart of a resolution logic $\mathcal{P}$. We say that a variable $p$ is *positive* in a formula $\alpha$ if

(p+) for every $\beta_0, \beta_1 \in L^*, \beta_0 \leq_{Rs} \beta_1$ implies $\alpha(p/\beta_0) \leq_{Rs} \alpha(p/\beta_1)$.

We say that $p$ is *negative* in $\alpha$ if

(p−) for every $\beta_0, \beta_1 \in L^*, \beta_0 \leq_{Rs} \beta_1$ implies $\alpha(p/\beta_1) \leq_{Rs} \alpha(p/\beta_0)$.

$p$ is said to be of *no polarity* in $\alpha$ if it is neither positive nor negative. We call this notion of polarity the *unrestricted polarity*.

THEOREM 5.10: *The assignment of unrestricted polarity values to variables occurring in formulas of $L^*$ can be done effectively.*

*Proof:* To assign the polarity values to variables in formulas effectively we can use (p+) and (p−) as tests with '$L^*$' replaced by '$Ver_{Rs}$'. To prove this claim, let us consider (p+). Suppose that $\alpha(p), \beta_0, \beta_1 \in L^*$ are such that

$$\beta_0 \leq_{Rs} \beta_1 \text{ but } \alpha(p/\beta_0) \leq_{Rs} \alpha(p/\beta_1) \text{ does not hold.}$$

We must find two verifiers $v_0$ and $v_1$ such that $v_0 \leq_{Rs} v_1$ but $\alpha(p/v_0) \leq_{Rs} \alpha(p/v_1)$ does not hold. To this end, let $e$ be an interpretation substitution and let $d$ be a set of designated truth-values of $M_{Rs}$ such that

$$e^*(\alpha(p/\beta_0)) \in d \text{ while } e^*(\alpha(p/\beta_1)) \notin d.$$

Since $e^*(\alpha(p/e^*(\beta_0))) = e^*(\alpha(p/\beta_0))$ and $e^*(\alpha(p/e^*(\beta_1))) = e^*(\alpha(p/\beta_1))$, we also have

(a) $e^*(\alpha(p/e^*(\beta_0))) \in d$ while $e^*(\alpha(p/e^*(\beta_1))) \notin d$.

Clearly, $\beta_0 \leq_{Rs} \beta_1$ implies

(b) $e^*(\beta_0) \leq_{Rs} e^*(\beta_1)$.

Since $e^*(\beta_0)$ and $e^*(\beta_1)$ are verifiers of $Rs$, (a) and (b) justify the claim for (p+). In the case of (p−) the proof is similar. ∎

THEOREM 5.11: *Rs admits* (pol1) *formulated for the unrestricted polarity.*

*Proof:* The proof of this theorem is analogous to that of Theorem 5.8. The only modification is that the definitions (p+) and (p−) play the role of Theorem 5.7. ∎

EXAMPLE 5.12: Let us return to our test sets $X_0$ and $X_1$. We know, from Example 5.9, that (pol1) formulated for the operator polarity blocks 58 initial applications of the resolution rule to the formulas of both sets. Although the unrestricted polarity fails to assign polarity values to $p_1$ and $p_2$ in the formulas of $X_0$, it makes $p_2$ negative in $\sim p_2$ causing (pol1) to block all but 76 initial applications of the resolution rule of $Rs_3$ to $X_1$ (cf. Table 5.1 in Section 5.9).

∎

Example 5.12 shows that (pol1) defined for the unrestricted polarity is more powerful than when formulated for the operator polarity. However, unless further polarity assignment algorithms more efficient than the test that follows from the proof of Theorem 5.10 are found, the computation of unrestricted polarity values can be quite costly. In practice, we may want to trade the precision of computing polarity values for better time complexity. The notion of the operator polarity is one of possible compromises; other solutions are presented in Sections 5.5 and 5.8.

## 5.5 VERIFIER POLARITY

The reason why the unrestricted polarity blocks more initial applications of the resolution rule of $Rs_3$ to the set $X_1$ than to $X_0$ is the fact that $p_2$ is negative in $\sim p_2$, but it has no polarity in $\neg p_2$. (pol1) can be made more powerful if, in its definition, the polarity relation $\leq_{Rs}$ is replaced with a weaker relation $\prec$, and if $\prec$ is used as the basis for defining the notion of a polarity value. For many resolution logics, including $\mathcal{P}_3$, such a modified version of polarity does a better job in assigning polarity values and triggers (pol1) in many cases when the unrestricted polarity keeps (pol1) idle. Now, the details.

Let $\prec$ be a partial order of the set $Ver$ of verifiers of $Rs$ that is consistent with $\leq_{Rs}$, i.e., $\prec$ is a subrelation of $\leq_{Rs}$ restricted to $Ver$. In Figure 5.3 we depict one of such orderings; $v_i \prec v_j$ if and only if there is a directed path from $v_i$ to $v_j$.

We say that a variable $p$ is *positive* in a formula $\alpha \in L^*$ with respect to $\prec$ if

(vp+) for every $v, w \in Ver$, $v \prec w$ implies $\alpha(p/v) \leq_{Rs} \alpha(p/w)$.

We say that $p$ is *negative* in $\alpha \in L^*$ with respect to $\prec$ if

(vp−) for every $v, w \in Ver$, $v \prec w$ implies $\alpha(p/w) \leq_{Rs} \alpha(p/v)$.

$p$ is said to be of *no polarity* in $\alpha$ if it is neither positive nor negative. We call this notion of polarity the *verifier polarity*.

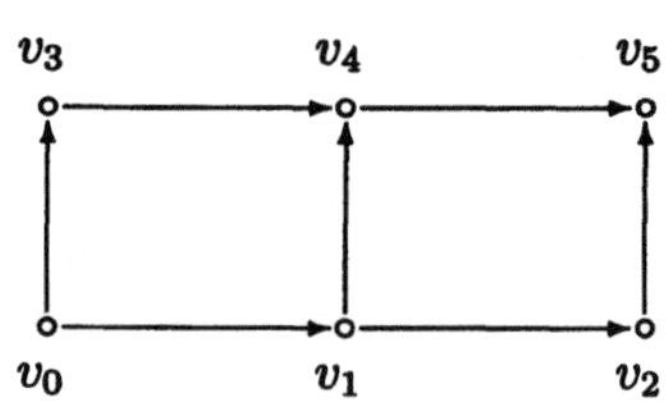


FIG. 5.3: Relation $\prec$ for $Rs_3$.

From the definitions of ($vp+$) and ($vp-$) it follows that the polarity values of variables can be effectively assigned.

EXAMPLE 5.13: Let us return to the system $Rs_3$. If $\prec$ is the partial order of the set of verifiers of $Rs_3$ as graphed in Figure 5.3, then for every variable $p$:

(a) $p$ is positive in itself, in $p \vee \beta$, $\beta \vee p$, $p \wedge \beta$, $\beta \wedge p$, and in $\beta \rightarrow p$, for every $\beta \in L^*$;

(b) $p$ is negative in $\neg p$, in $\sim p$, and in $p \rightarrow \beta$, for every $\beta \in L^*$.

From (a) and (b) we conclude that, in contrast to the notions of polarity discussed in the preceding sections, the variables $p_1, p_2$, and $p_3$ are '+' or '−' in every formula of $X_0 \cup X_1$. ∎

THEOREM 5.14: *Rs admits* (pol1) *formulated for the verifier polarity.*

*Proof:* The proof of this theorem is analogous to that of Theorem 5.8. The only modification is that $\prec$ replaces $\leq_{Rs}$ in (pol1) and that the definitions (vp+) and (vp−) play the role of Theorem 5.7. ∎

EXAMPLE 5.15: (pol1) formulated for the unrestricted polarity blocks 58 initial applications of the resolution rule of $Rs_3$ to the set $X_0$ and 116 applications of this rule to the formulas of $X_1$. Using the facts reported in Example 5.13 we conclude that (pol1) formulated for the verifier polarity blocks all but 27 initial applications of the resolution rule to the formulas of both $X_0$ and $X_1$ (cf. Table 5.1 in Section 5.9). ∎

## 5.6 STRENGTHENING THE POLARITY STRATEGY

Let $\mathcal{P}$, $Rs$, and $M_{Rs}$ have the same meanings as in the preceding sections, and let us assume that among the verifiers of $Rs$ there is a tautological verifier $v_T$ (i.e., $v_T \in C(\emptyset)$) and a contradictory verifier $v_F$ (i.e., $C(v_F) = L$). Let us strengthen the polarity strategy (pol1) in the following way.

(pol2) *The resolvent* $\beta_0(p/v_0) \vee \ldots \vee \beta_n(p/v_n)$ *of* $\beta_0(p), \ldots, \beta_n(p)$ *cannot be formed and used during the deductive process if its formation is blocked by* (pol1), *if* $p$ *is only positive in every* $\beta_i$, *or, finally, if* $p$ *is only negative in every* $\beta_i$.

In other words, we can form and use the resolvent $\beta_0(p/v_0) \vee \ldots \vee \beta_n(p/v_n)$ if its formation is allowed by (pol1) and, in addition, if either $p$ has no polarity in some $\beta_i$ or $p$ is positive in some $\beta_i$ and negative in some $\beta_j$.

In this section we show that $Rs$ admits (pol2) formulated for all three variants of polarity provided that the inference engine of $Rs$ is supplemented with two new inference rules:

$$R^+ : \frac{\alpha(p^\oplus)}{\alpha(p/v_T)}, \qquad R^- : \frac{\alpha(p^\ominus)}{\alpha(p/v_F)}.$$

We read '$p^\oplus$' and '$p^\ominus$' as '$p$ is only positive in $\alpha$' and '$p$ is only negative in $\alpha$', respectively.

We say that $Rs$ admits the polarity strategy (pol2), if this system remains refutationally complete for $\mathcal{P}$ while its resolution rule is restricted by (pol2) and its set of inference rules is expanded by the addition of $R^+$ and $R^-$.

LEMMA 5.16: *Let $X \subseteq L$ be finite and clean in $Rs$. If there exists a* (pol2)-*restricted refutation of $X$ in $Rs$, then $X$ is $C$-inconsistent.*

*Proof:* Let $X$ be as stated and let

(a) $\alpha_0, \ldots, \alpha_k, \beta_0, \ldots, \beta_l$

be a (pol2)-restricted refutation of $X$. To make our considerations less complex, let us assume that $\alpha_k$ and $\beta_l$ are verifiers and that $\{\alpha_k, \beta_l\} \in \mathcal{F}$. Moreover, let us assume that $\alpha_0, \ldots, \alpha_k$ is the derivation of $\alpha_k$ and that $\beta_0, \ldots, \beta_l$ is the derivation of $\beta_l$ from $X$. Our objective is to transform (a) into a refutation of $X$ in $Rs$ which requires no applications of $R^+$ or $R^-$. To this end, let us consider the sequence $\alpha_0, \ldots, \alpha_k$. Without any loss of generality we can assume that no application of a transformation rule precedes an application of the resolution rule, $R^+$, or $R^-$.

Let $\alpha_0^*, \ldots, \alpha_k^*$ be a sequence of formulas satisfying the following properties: for every $j \leq k$,

(b) $\alpha_j$ and $\alpha_j^*$ have the same variables;

(c) $\alpha_j$ =[$Rs$] $\alpha_j^*$.

The sequence in question is defined as follows. If $\alpha_i \in X$, then $\alpha_i^* = \alpha_i$. Let us suppose that for some $j \geq 0, \alpha_0^*, \ldots, \alpha_j^*$ has been constructed. Let us consider $\alpha_{j+1}$. If $\alpha_{j+1} \in X$, then $\alpha_{j+1}^*$ has already been defined. If this formula is the resolvent of some $\alpha_{i_0}, \ldots, \alpha_{i_n}$ upon $p$, then we let $\alpha_{j+1}^*$ be the resolvent of $\alpha_{i_0}^*, \ldots, \alpha_{i_n}^*$ upon $p$. Clearly, (b) holds. The truth of (c) follows from the inductive hypothesis and Exercise 5.1. If $\alpha_{j+1}$ is the result of the application of $R^+$ or $R^-$ to $\alpha_i(p)$, then let $\alpha_{j+1}^* = \alpha_i^*(p/v_0) \vee \ldots \vee \alpha_i^*(p/v_n)$. Again, it is easy to see that $\alpha_{j+1}$ and $\alpha_{j+1}^*$ satisfy (b) and that Exercise 5.1, together with the inductive hypothesis, can be used to justify (c).

Now, let $\alpha_j$ be the first formula to which a transformation rule has been applied. Moreover, let $\alpha_{j+1}^*, \ldots, \alpha_k^*$ be the sequence of the successive simplifications of $\alpha_j^*$ obtained with the help of the transformation rules of $Rs$. (In general to transform $\alpha_j^*$ into a verifier may require more than $k - j$ transformations. The assumption that $\alpha_k^*$ is a verifier has been made to simplify the proof.) By (r2) and (c), for every set $d$ of designated truth-values of $M_{Rs}$,

(d) $\alpha_k \in d$ *iff* $\alpha_k^* \in d$.

This finishes the construction of the sequence $\alpha_0^*, \ldots, \alpha_k^*$.

In the same way we construct the sequence $\beta_0^*, \ldots, \beta_l^*$ and show that (d) holds for $\beta_l$ and $\beta_l^*$. Clearly, $\alpha_0^*, \ldots, \alpha_k^*, \beta_0^*, \ldots, \beta_l^*$ is a deduction of $\{\alpha_k^*, \beta_l^*\}$ from $X$ in $Rs$. By (d), this set is an element of $\mathcal{F}$ and our proof is finished. ∎

THEOREM 5.17: *$Rs$ admits* (pol2) *formulated for each of the variants of polarity.*

*Proof:* The proof of this theorem for the operator polarity can be found in [43].

For the unrestricted polarity we proceed as follows. Let $X \subseteq L$ be finite and clean in $Rs$. By Lemma 5.16, it suffices to show that $C(X) = L$ implies the existence of a (pol2)-restricted refutation of $X$ in $Rs$. Let us reconsider the proof of Theorem 5.8. To make that proof work under the current assumptions, we have to modify only that part of the proof which is concerned with the selection of a formula falsified by the inference node $K$.

Let $\alpha_0(p_{k+1}), \ldots, \alpha_n(p_{k+1})$ be formulas falsified by the children of $K$. If $p_{k+1}$ is only positive in every $\alpha_i$, then we simulate the application of $R^+$ to $\alpha_n(p_{k+1})$. Clearly, the partial valuation associated with $K$ falsifies $\alpha_n(p/p_{v_T})$. If $p_{k+1}$ is only negative in every $\alpha_i$, then we simulate $R^-$ to generate $\alpha_0(p_{k+1}/p_{v_F})$. Again, the partial valuation associated with $K$ falsifies this formula. If neither of these cases apply to the formulas $\alpha_0, \ldots, \alpha_n$, (pol2) is satisfied and we can use the resolution rule to turn $K$ into a failure node.

For the verifier polarity, the proof of this theorem is similar. ∎

EXAMPLE 5.18: Before we return to our test sets $X_0$ and $X_1$, let us demonstrate how the rules $R^+$ and $R^-$ can be used in the deductive process. Let $\alpha = \neg(p_1 \to (p_2 \to p_1))$. Refuting $\{\alpha\}$ in $Rs_2$ without $R^+$ and $R^-$, requires two applications of the resolution rule of $Rs_2$ and fifteen applications of transformation rules. However, $\{\alpha\}$ can be refuted in $Rs_2$ using just one application of the resolution rule, one application of $R^+$, and only 7 applications of transformation rules:

(1) $\neg(p_1 \to (T \to p_1))$, ($R^+$ applied to $p_2$ in $\alpha$)

(2) $\neg(F \to (T \to F)) \vee \neg(T \to (T \to T))$, (resolvent of (1), (1))

(3) $F$. (transformation of (2))

Now, back to the test sets $X_0$ and $X_1$. Let $v_F$ be $v_0$ and $v_T$ be $v_5$. In addition to the initial applications of the resolution rule of $Rs_3$ blocked by (pol1), (pol2) does not allow the resolution of any sequence of identical formulas upon a variable which is positive or negative in these formulas. As a result, (pol2) formulated for the verifier polarity blocks all but 21 initial applications of the resolution rule to the formulas of $X_0$ and $X_1$ (cf. Table 5.1 in Section 5.9). ∎

## 5.7 POLARITY AND THE SIMPLIFICATION OF RESOLVENTS

In the preceding section we studied the effect of (po1) and (po2) on the reduction of the size of the search space. These strategies can also be used to control the length of resolvents, a fact invisible in the case of classical propositional logic but fully transparent in cases that require more than two verifiers.

We begin with an example that illustrates the key features of our method of reducing the length of resolvents using polarity.

EXAMPLE 5.19: Let us consider the unrestricted polarity and the test set

$$X_2 = \{p_3, p_3 \supset p_1, p_1 \supset p_2, \sim p_2\},$$

where '$q \supset r$' abbreviates '$\sim q \vee r$'. Let us note that in $Rs_3$, $r$ is positive in $q \supset r$ while $q$ is negative in this formula. One of the sequences of formulas of $X_2$ which can be resolved in $Rs_3$ without violating (pol2) is

(a) $$p_1 \supset p_2,\ p_1 \supset p_2,\ \sim p_2,\ p_1 \supset p_2,\ p_1 \supset p_2,\ \sim p_2.$$

The resolvent of (a) is the formula

(b) $$\alpha(p_1) = (p_1 \supset v_0) \vee (p_1 \supset v_1) \vee \sim v_2 \vee (p_1 \supset v_3) \vee (p_1 \supset v_4) \vee \sim v_5.$$

Since $v_0 =[Rs_3]\ v_1$ and since $p_2$ is positive in $p_1 \supset p_2$, for every interpretation substitution $e$ and every set $d$ of designated truth-values of the matrix $M$ induced by $Rs_3$,

$$e^*(p_1 \supset v_0) \in d \quad \textit{iff} \quad e^*(p_1 \supset v_1) \in d.$$

Since $M$ is well-connected (cf. Lemma 3.10), from (b) we get

$$e^*(\alpha(p_1)) \in d \quad \textit{iff} \quad e^*((p_1 \supset v_0) \vee \sim v_2 \vee (p_1 \supset v_3) \vee (p_1 \supset v_4) \vee \sim v_5) \in d.$$

Since also $v_3 =[Rs_3]\ v_4$,

(c) $$e^*(\alpha(p_1)) \in d \quad \textit{iff} \quad e^*((p_1 \supset v_0) \vee \sim v_2 \vee (p_1 \supset v_3) \vee \sim v_5) \in d.$$

The derivation of (c) suggests how to simplify the resolution rule of $Rs_3$. Let us suppose that $\alpha_0(p), \ldots, \alpha_5(p)$ is a sequence of formulas which satisfies (pol2) with respect to the variable $p$. If $p$ has a polarity value in $\alpha_0$ and $\alpha_3$, then the resolvent $\alpha_0(p/v_0) \vee \ldots \vee \alpha_5(p/v_5)$ can be shortened by eliminating the second and the fifth disjunct. Hence, to form the resolvent of (a) we require not all but just four verifiers and (b) can be rewritten as

$$(p_1 \supset v_0) \vee \sim v_2 \vee (p_1 \supset v_3) \vee \sim v_5.$$

■

Now, let us generalize the observations made in the above example. The *polarized resolution rule* for $Rs$ is the set of all sequents of the form

$$R_{pol}: \frac{\alpha_0(p), \ldots, \alpha_n(p)}{\alpha_{i_0}(p/v_{i_0}) \vee \ldots \vee \alpha_{i_k}(p/v_{i_k})}.$$

In this sequent, $p$ is a variable that does not occur in the verifiers of $Rs$ and the sequence $\alpha_0(p), \ldots, \alpha_n(p)$ satisfies (pol2) with respect to $p$. The conclusion $\alpha_{i_0}(p/v_{i_0}) \vee \ldots \vee \alpha_{i_k}(p/v_{i_k})$ (called the *$R_{pol}$-resolvent* of $\alpha_0, \ldots, \alpha_n$) is obtained from the resolvent $\alpha_0(p/v_0) \vee \ldots \vee \alpha_n(p/v_n)$ by removing every disjunct $\alpha_j(p/v_j)$ such that for some $i < j$, $v_i =[Rs]\ v_j$ and $p$ is positive or negative in $\alpha_i$.

THEOREM 5.20: *Let $Rs$ be a resolution counterpart of a resolution logic $\mathcal{P}$. Then $Rs$ with the resolution rule replaced by $R_{pol}$ is refutationally complete for $\mathcal{P}$, i.e., $Rs$ and $\mathcal{P}$ satisfy* (r1).

*Proof:* We present the proof for the unrestricted polarity. For the operator and verifier variants of polarity the proofs are similar.

Let $X \subseteq L$ be finite, clean in $Rs$, and $C$-inconsistent. Let us look at the proof of Theorem 5.8. To make this proof work under the current assumptions, we have to modify only that part of the proof which is concerned with the choice of a formula falsified by the inference node $K$.

Let $\alpha_0(p_{k+1}), \ldots, \alpha_n(p_{k+1})$ be formulas falsified by the children of $K$ and let us assume that for some $i < j \leq n, v_i =[Rs]\ v_j$ and that $p_{k+1}$ is positive in $\alpha_i$. Let $h$ be the partial valuation associated with $K$, and let $e$ be any interpretation substitution such that for every variable $p$, $e(p) = h(p)$, if $h(p)$ is defined. So,

(a) $e^*(\alpha_l(p_{k+1}/v_l)) \notin d$, for every $l \leq n$.

Since $v_i =[Rs]\ v_j$ and since $p_{k+1}$ is positive in $\alpha_i$,

(b) $e^*(\alpha_i(p_{k+1}/v_i)) \in d$ *iff* $e^*(\alpha_i(p_{k+1}/v_j)) \in d$.

(a), (b), and Exercise 5.1 imply that $K$ falsifies the formula $\alpha = \alpha_0(p_{k+1}/p_{v_0}) \vee \ldots \vee \alpha_{j-1}(p_{k+1}/p_{v_{j-1}}) \vee \alpha_{j+1}(p_{k+1}/p_{v_{j+1}}) \vee \ldots \vee \alpha_n(p_{k+1}/p_{v_n})$. If $p_{k+1}$ is negative in $\alpha_i$ we proceed in the same way. If there are other pairs $v_s, v_t$ of verifiers such that:

$s < t \leq n, v_s =[Rs]\ v_t$, and $p$ is positive or negative in $\alpha_s$, then $\alpha$ can be further simplified (in the way described above) and will eventually result in a formula that corresponds to the $R_{pol}$-resolvent of $\alpha_0, \ldots, \alpha_n$ and which will be falsified by $K$.

The proof that finite, clean in $Rs$, and $C$-consistent sets cannot be refuted in $Rs$ with $R^{\vee}_{Ver}$ replaced by $R_{pol}$, is left to the reader as an exercise. ∎

In the light of Theorem 5.20, many resolvents can be formed during the deductive process in $Rs_3$ using only 4 verifiers. This is because $v_0 =[Rs_3]\ v_1$ and $v_3 =[Rs_3]\ v_4$. The reduction of the number of verifiers used in the formation of resolvents is even more drastic in the case of other $n$-valued Łukasiewicz logics, $n > 2$. Although the minimal resolution counterpart $Rs_n$ of $\mathcal{P}_n$ has $2n$ verifiers, by Exercise 5.2, the relation $=[Rs_n]$ divides the set of verifiers of $Rs_n$ into exactly 4 equivalence classes. Hence, still, only 4 verifiers are required to resolve a sequence of formulas in $Rs_n$ upon a variable which, e.g., has a polarity value in every formula of the sequence.

EXAMPLE 5.21: The set $X_2$, defined in Example 5.19, is inconsistent in $\mathcal{P}_3$ and one of its refutations in $Rs_3$ can be constructed using $R_{pol}$ in the following way. First, we apply $R_{pol}$ to the sequence

$$p_1 \supset p_2,\ \ p_1 \supset p_2,\ \ \sim p_2,\ \ p_1 \supset p_2,\ \ p_1 \supset p_2,\ \ \sim p_2$$

and obtain $(p_1 \supset v_0) \vee \sim v_2 \vee (p_1 \supset v_3) \vee \sim v_5$, which reduces to the formula

$$\gamma(p_1) = (p_1 \supset v_0) \vee (p_1 \supset v_3) \vee v_3.$$

Next, we apply the resolution rule to the sequence

$$p_3 \supset p_1,\ \ p_3 \supset p_1,\ \ \gamma(p_1),\ \ p_3 \supset p_1,\ \ p_3 \supset p_1,\ \ \gamma(p_1)$$

and obtain $(p_3 \supset v_0) \vee \gamma(p_1/v_2) \vee (p_3 \supset v_3) \vee \gamma(p_1/v_5)$, which reduces to the formula

$$\delta(p_3) = (p_3 \supset v_0) \vee (p_3 \supset v_3) \vee v_4.$$

Finally, we apply the resolution rule to:

$$p_3,\ \ p_3,\ \ \delta(p_3),\ \ p_3,\ \ p_3,\ \ \delta(p_3)$$

and obtain $v_0 \vee \delta(p_3/v_2) \vee v_3 \vee \delta(p_3/v_5)$, which reduces to $v_4$. By Exercise 5.4 and condition (T2*) formulated in Example 4.24, the refutation of $X_2$ terminates successfully. ∎

## 5.8 POLARITY AND THE DEDUCTIVE PROCESS

In the preceding sections we introduced three variants of the notion of polarity and discussed the restrictive power of polarity strategies (pol1) and (pol2) formulated for all these variants. In this section we briefly discuss possible ways in which these techniques can be used in an automated proof system.

THEOREM 5.22: *Let $Rs$ be a resolution proof system on $\mathcal{L}$ and let $\alpha(p)$ be an arbitrary formula of $L^*$. Then:*

(i) *if $p$ is '$+$' (is '$-$') in $\alpha$ with respect to the operator polarity, then it is '$+$' (is '$-$') in $\alpha$ with respect to the unrestricted polarity;*

(ii) *if $p$ is '$+$' (is '$-$') in $\alpha$ with respect to the unrestricted polarity, then it is '$+$' (is '$-$') in $\alpha$ with respect to the verifier polarity.*

*Proof:* (i) follows from Theorem 5.7. (ii) follows from the fact that $\prec$ is a subrelation of $\leq_{Rs}$ restricted to $Ver_{Rs}$. ∎

Theorem 5.22 shows that the unrestricted polarity does a better job in assigning polarity values than the operator polarity, and that the verifier polarity is superior to the unrestricted polarity. Hence, in general, the verifier polarity will trigger (pol2) more often than the unrestricted polarity. In some cases, however, (pol2) formulated for the unrestricted polarity can have a greater effect on the reduction of the search space than when formulated for the verifier polarity. For example, there are 7 initial applications of the resolution rule of $Rs_3$ to the formulas of $\{\sim p_2, p_1 \rightarrow p_2\}$ when the rule is restricted by (pol2) formulated for the verifier polarity, while there are only 4 such applications when (pol2) is taken with respect to the unrestricted polarity. The reason is that in the case of the verifier polarity, (pol1) is defined not with respect to the whole polarity relation $\leq_{Rs}$ but only with respect to its subrelation $\prec$.

From the practical point of view, one would like to develop the notion of polarity that acts with the precision of the verifier polarity while retaining the algorithmic simplicity exhibited by the operator polarity assignment algorithm. The notion of the *verifier-operator polarity* (*vo-polarity*), which we describe below, is such an attempt. To define this notion we first select an ordering $\prec$ of the set of verifiers as in the case of the verifier polarity. Then, we accept the definitions (op+) and (op$-$) of a functor being positive or negative over its arguments with the following changes. The symbol '$L^*$' is replaced by '$Ver_{Rs}$' and the first occurrence of '$\leq_{Rs}$' in both (op+) and (op$-$) is replaced by '$\prec$'. The notions of a *positive* (resp. *negative*) occurrence of a formula as well as the notions of a variable being *positive* (resp. *negative*) in a formula are defined in the same way as in Section 5.3.

THEOREM 5.23: *Let $Rs$ be a resolution counterpart of $\langle \mathcal{L}, C \rangle$. Then $Rs$ admits* (pol2) *formulated for the vo-polarity.*

*Proof:* Since Lemma 5.16 holds for *vo*-polarity, we only need to justify that every finite, clean, and $C$-inconsistent set of formulas can be refuted in $Rs$, when the resolution rule of $Rs$ is restricted by (pol2) formulated for *vo*-polarity. To this end, we simply repeat the proof of Theorem 5.8. The only modification required is the replacement of Theorem 5.7 by the following two facts concerning *vo*-polarity. Let $\alpha(p) \in L^*$ and let $v, w \in Ver$ be such that $v \prec w$. Then:

(a) if $p$ is positive in $\alpha(p)$, then $\alpha(p/v) \leq_{Rs} \alpha(p/w)$;

(b) if $p$ is negative in $\alpha(p)$, then $\alpha(p/w) \leq_{Rs} \alpha(p/v)$.

(a) and (b) can be easily demonstrated by induction on the complexity of $\alpha$. ∎

The notion of *vo*-polarity is stronger than the notion of the operator polarity but weaker than that of the verifier polarity. For instance, the verifier polarity assigns '+' to $p$ in $(v_2 \to p) \to p$ in $Rs_3$, while $p$ is neither positive nor negative in this formula with respect to *vo*-polarity. However, every connective of $\mathcal{P}_3$ has a polarity value over all its arguments with respect to *vo*-polarity. As a result, the number of the initial applications of the resolution rule to the test sets $X_0$ and $X_1$ is the same for both the verifier and *vo*-polarities (see Table 5.1 in the next section).

The experiments with the test sets $X_0$ and $X_1$ demonstrate that the restrictive power of (pol2) increases when this restriction condition is simultaneously verified with respect to two or more variants of polarity. Hence, if possible, all the variants of polarity discussed in this chapter should be implemented in a resolution based automated reasoning system. The effect of the set of support strategy on the number of initial applications of the resolution rule to the formulas of $X_0$ and $X_1$ is also evident (see Table 5.1 in the next section).

Our last theorem of this chapter shows that the combination of the polarity and set of support strategies preserves refutational completeness of resolution counterparts of propositional logics.

THEOREM 5.24: *Let $Rs$ be a resolution counterpart of a logic $\langle \mathcal{L}, C \rangle$, let $X \subseteq L$ be finite and clean in $Rs$, and let $Y$ be a set of support of $X$. Then $C(X) = L$ iff there exists a* (pol2)*-restricted, $(X, Y)$-supported refutation of $X$ in $Rs$.*

*Proof:* Let $C, Rs, X$, and $Y$ be as stated. If there exists a (pol2)-restricted, $(X, Y)$-supported refutation of $X$ in $Rs$, then, by Lemma 5.16, $C(X) = L$.

If $C(X) = L$, then we repeat the proof of Theorem 5.1 with the following modification concerning the choice of a formula falsified by the inference node $K$. Let $\alpha_0(p_{k+1}), \ldots, \alpha_n(p_{k+1})$ be formulas falsified by the children of $K$. If $p_{k+1}$ is positive in every $\alpha_i$, then we simulate the application of $R^+$ to $\alpha_n(p_{k+1})$. Clearly, the partial valuation associated with $K$ falsifies $\alpha_n(p/p_{v_T})$. If $p_{k+1}$ is negative in every $\alpha_i$, then we simulate $R^-$ to generate $\alpha_0(p_{k+1}/p_{v_F})$. Again, the partial valuation associated with $K$ falsifies this formula. If neither of these cases apply to the formulas $\alpha_0, \ldots, \alpha_n$, then we can assume that

(a) $\alpha_0(p_{k+1}), \ldots, \alpha_n(p_{k+1})$ satisfy (pol1) and for some $j \leq n, \alpha_j(p_{k+1}) \notin X - Y$.

By Theorems 5.8, 5.11, 5.14, and 5.23, the satisfaction of the first part of (a) can always be guaranteed. The satisfaction of the second part of (a) can be demonstrated as in the proof of Theorem 5.1. To conclude, if $C(X) = L$, then there exists a (pol2)-restricted and $(X, Y)$-supported refutation of $X$. ∎

## 5.9 TEST RESULTS

Let $Rs_3$ be the resolution counterpart of the 3-valued Łukasiewicz logic $\mathcal{P}_3$ (cf. Appendix B). The following two sets of formulas of $\mathcal{P}_3$ have been extensively used in the preceding sections:

$$X_0 = \{p_3, p_3 \to p_1, p_1 \to p_2, \neg p_2\},$$
$$X_1 = \{p_3, p_3 \to p_1, p_1 \to p_2, \sim p_2\},$$

where '$\sim p_2$' abbreviates '$\neg p_2 \vee ((p_2 \to \neg p_2) \wedge (\neg p_2 \to p_2))$'.

In Table 5.1, we have collected information concerning the effect of the restriction strategies discussed in this chapter on the number of initial applications of the resolution rule of $Rs_3$ to the formulas of $X_0$ and $X_1$. A row of the table shows the number of initial applications of the resolution rule when an indicated strategy, or a combination of strategies, is used.

TABLE 5.1: The number of initial applications of the resolution rule of $Rs_3$ to the formulas of $X_0$ and $X_1$.

| strategies: | number of applications of the resolution rule of $Rs_3$ to: $X_0$ | $X_1$ |
|---|---|---|
| none | 192 | 192 |
| (pol1) operator | 134 | 134 |
| (pol2) operator | 133 | 133 |
| (pol1) unrestricted | 134 | 76 |
| (pol2) unrestricted | 133 | 74 |
| (pol1) verifier | 27 | 27 |
| (pol2) verifier | 21 | 21 |
| (pol1) *vo* | 27 | 27 |
| (pol2) *vo* | 21 | 21 |
| (pol2) unrestricted + verifier | 18 | 15 |
| (pol2) *vo* + set of support | 7 | 7 |
| (pol2) unrestricted + verifier + set of support | 7 | 4 |

We can start a refutation of $X_0$ in $Rs_3$ in 192 different ways. One of the $(X, \{\neg p_2\})$-supported refutations of $X_0$ can be constructed by repeating the steps of the refutation constructed in Example 5.21. First, we apply the resolution rule to the sequence

$$p_1 \to p_2^{\oplus},\ p_1 \to p_2^{\oplus},\ \neg p_2^{\ominus},\ p_1 \to p_2^{\oplus},\ p_1 \to p_2^{\oplus},\ \neg p_2^{\ominus}$$

and obtain $(p_1 \to v_0) \vee (p_1 \to v_1) \vee \neg v_2 \vee (p_1 \to v_3) \vee (p_1 \to v_4) \vee \neg v_5$, which reduces to the formula

$$\gamma(p_1) = (p_1 \to v_0) \vee (p_1 \to v_1) \vee (p_1 \to v_3) \vee (p_1 \to v_4) \vee v_3.$$

(The superscripts $\oplus$ and $\ominus$ indicate the positive and negative verifier polarity values.) Next, we apply the resolution rule to the sequence

$$p_3 \to p_1^{\oplus},\ p_3 \to p_1^{\oplus},\ \gamma(p_1^{\ominus}),\ p_3 \to p_1^{\oplus},\ p_3 \to p_1^{\oplus},\ \gamma(p_1^{\ominus})$$

and obtain $(p_3 \to v_0) \vee (p_3 \to v_1) \vee \gamma(p_1/v_2) \vee (p_3 \to v_3) \vee (p_3 \to v_4) \vee \gamma(p_1/v_5)$, which reduces to the formula

$$\delta(p_3) = (p_3 \to v_0) \vee (p_3 \to v_1) \vee (p_3 \to v_3) \vee (p_3 \to v_4) \vee v_4.$$

Finally, we apply the resolution rule to the sequence

$$p_3^{\oplus},\ p_3^{\oplus},\ \delta(p_3^{\ominus}),\ p_3^{\oplus},\ p_3^{\oplus},\ \delta(p_3^{\ominus})$$

and obtain $v_0 \vee v_1 \vee \delta(p_3/v_2) \vee v_3 \vee v_4 \vee \delta(p_3/v_5)$, which reduces to $v_4$. In view of (T2*), formulated in Example 4.24, and Exercises 5.4 and 5.5, the refutation terminates successfully.

Incidentally, the same sequence of applications of the resolution rule will lead us to a refutation of $X_1$ in $Rs_3$.

**EXERCISES**

5.1. Let $Rs$ be a resolution counterpart of a disjunctive logic. Moreover, let $\alpha, \beta \in L^*$, let $e$ be an interpretation substitution, and let $d$ be a set of designated truth-values of $M_{Rs}$. Show that $e^*(\alpha \vee \beta) \in d$ if and only if $e^*(\alpha) \in d$ or $e^*(\beta) \in d$.

5.2. Let $n > 2$ and let $Rs_n$ be the resolution counterpart of the $n$-valued Łukasiewicz logic constructed in Section 4.3. Show that the equivalence relation $=[Rs_n]$ divides the set of verifiers of $Rs_n$ into exactly four equivalence classes.

5.3. Let $Rs$ be a resolution counterpart of a resolution logic $\langle \mathcal{L}, C \rangle$, let $v_F$ be a contradictory verifier (i.e., $C(v_F) = L$), and let $v_T$ be a tautological verifier (i.e., $v_T \in C(\emptyset)$). Show that for every $v \in Ver$, $v_F \leq_{Rs} v \leq_{Rs} v_T$.

5.4. Let $M$ be as in Theorem 4.17 and let $Rs$ be a resolution counterpart of $\langle \mathcal{L}, C_M \rangle$. For every finite and clean $X \subseteq L$, let $r_{pol}(X)$ be the set of all verifiers of $Rs$ that can be deduced from $X$ using: $R^+, R^-$, the transformation rules of $Rs$, and the resolution rule of $Rs$ restricted by (pol2) formulated for the operator or unrestricted polarity. Show that we can select a maximal consistent set $d$ of verifiers of $Rs$ in such a way that for every clean and finite $X \subseteq L$,

$$C_M(X) = L \text{ iff } r_{pol}(X) \not\subseteq d.$$

5.5. Show that the fact stated in Exercise 5.4 remains true for the verifier and *vo* variants of polarity, provided that the relation $\prec$ is chosen so that for every verifier $v, v_F \prec v \prec v_T$, where $v_F$ and $v_T$ are as in Exercise 5.3.

5.6. Let $Rs, v_F$, and $v_T$ be as in Exercise 5.3. Let us assume that the application of $R^+$ to a formula $\alpha(p)$ is also allowed when $p$ is both positive and negative in $\alpha$. Let us strengthen (pol2) by blocking the formation of any resolvent $\beta_0(p/v_0) \vee \ldots \vee \beta_n(p/v_n)$ such that $p$ is both positive and negative in some $\beta_j$. Show that $Rs$ (with $R^+$ and $R^-$) admits this stronger version of (pol2) formulated for any of the variants of polarity.

5.7. Let $Rs$ be a resolution counterpart of a resolution logic. Recall that if $X \subseteq L$ is finite and clean, then $r(X)$ denotes the set of all resolvents that can be inferred from $X$ using the resolution rule of $Rs$. Let $\alpha(p)$ be the resolvent of $\alpha_0(p,q), \ldots, \alpha_n(p,q) \in r(X)$ upon $q$, and let $p \notin Var(Ver)$. Show that if $p$ is positive (is negative) in every $\alpha_i, i \leq n$, with respect to the verifier polarity, then it remains positive (it remains negative) in $\alpha(p)$. Show that this property holds for the remaining variants of polarity.

5.8. Suppose that during the deductive process in $Rs$ a formula $\alpha(p)$ is obtained from a formula $\beta(p)$ by the application of a transformation rule of $Rs$, where $p \notin Var(Ver)$. Show that if $p$ is positive (is negative) in $\beta(p)$ with respect to the verifier polarity, then it remains positive (it remains negative) in $\alpha(p)$. Show that this property holds for the remaining variants of polarity.

# 6

# RESOLUTION CIRCUITS

In Chapters 4 and 5 we discussed methods and strategies for an efficient implementation of the deductive process. In this chapter we continue this discussion; we introduce and study resolution circuits – multi-resolution proof system representations of resolution logics.

## 6.1 PROPOSITIONAL AND-OR CIRCUITS

One way of gaining efficiency of a resolution based reasoning program is a careful selection of a resolution proof system $Rs$ and the choice of appropriate strategies for directing and restricting the application of the inference rules of $Rs$. In some instances, we may expect further gains when we associate with a given logical system $\langle \mathcal{L}, C \rangle$ not a single resolution proof system but a finite class $\mathcal{R}$ of such systems. We may require, for instance, that for every finite $X \subseteq L$, clean in every system of $\mathcal{R}$,

$$C(X) = L \;\; \textit{iff} \;\; X \overset{Rs}{\Rightarrow} \Box, \text{ for every } Rs \in \mathcal{R}$$

(cf. Figure. 6.1). The systems of $\mathcal{R}$, when executed in parallel, determine the consistency status of a given set $X$ of formulas in time taken by the 'slowest' among the systems in question. We call the network of resolution proof systems shown in Figure 6.1 an AND resolution circuit. For many logical calculi this approach offers a substantial saving of computer time.

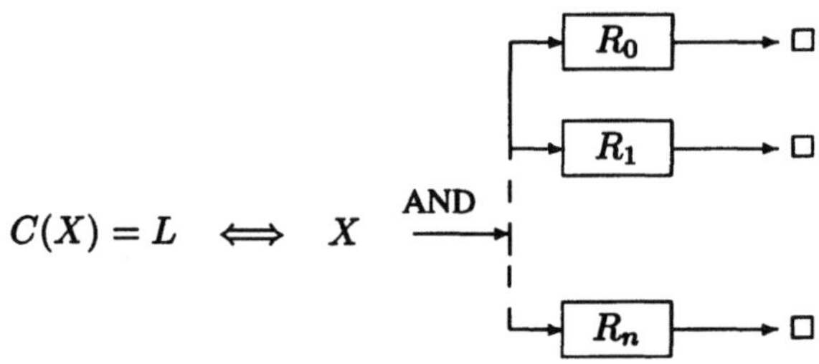


FIG. 6.1: AND resolution circuit for a logic $\langle \mathcal{L}, C \rangle$.

The notion of an AND circuit can be generalized to the notion of an AND-OR resolution circuit by, roughly speaking, allowing an arbitrary *and-or* connections between resolution proof systems. Let us look at the example pictured in Figure 6.2.

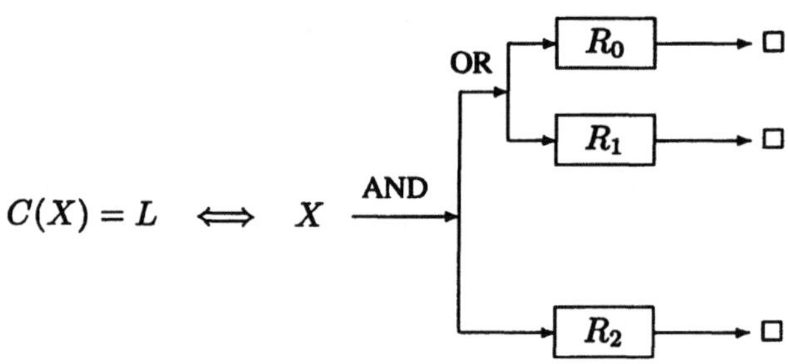


FIG. 6.2: AND-OR resolution circuit for a logic $\langle \mathcal{L}, C \rangle$.

The logic $\mathcal{P} = \langle \mathcal{L}, C \rangle$ is represented here by three resolution proof systems $R_0, R_1$, and $R_2$. For every finite set $X \subseteq L$, the consistency of $X$ in $\mathcal{P}$ is tested by running all these systems in parallel on input $X$. $X$ is inconsistent in $\mathcal{P}$ if and only if $R_2$ and one of the systems $R_0$ or $R_1$ refute $X$, that is,

$$X \textit{ is inconsistent in } \mathcal{P} \textit{ iff } X \overset{R_2}{\Rrightarrow} \Box \textit{ and } (X \overset{R_0}{\Rrightarrow} \Box \textit{ or } X \overset{R_1}{\Rrightarrow} \Box) \textit{ is true.}$$

The formula $\alpha(r_0, r_1, r_2) = r_2 \wedge (r_0 \vee r_1)$ of classical propositional logic represents the structure, or the type, of this resolution circuit. Using the type $\alpha$, the expression '$X \overset{R_2}{\Rightarrow} \square$ *and* $(X \overset{R_0}{\Rightarrow} \square$ *or* $X \overset{R_1}{\Rightarrow} \square)$' can be conveniently abbreviated as '$\alpha[r_i/X \overset{R_i}{\Rightarrow} \square]$', that is, $\alpha[r_i/X \overset{R_i}{\Rightarrow} \square]$ results from $\alpha$ by simultaneous replacement of every variable $r_i$ by $X \overset{R_i}{\Rightarrow} \square$. Clearly, $\alpha[r_i/X \overset{R_i}{\Rightarrow} \square]$ is a metalanguage formula and not a formula of $\mathcal{L}$.

Before we present the main results concerning resolution circuits, let us formally introduce the notion of an AND-OR resolution circuit of type $\alpha$. An *and-or formula* is a formula of classical propositional logic the only connectives of which are disjunction ($\vee$) and conjunction ($\wedge$). Let $\mathcal{P} = \langle \mathcal{L}, C \rangle$ be a resolution logic. By an AND-OR *resolution circuit* on $\mathcal{L}$ we understand a triple

$$\mathfrak{R} = \langle \mathcal{R}, \alpha, f \rangle,$$

where $\mathcal{R}$ is a finite set of resolution proof systems on $\mathcal{L}$, $\alpha$ is an *and-or* formula, and $f$ is an assignment of resolution proof systems of $\mathcal{R}$ to propositional variables of $\alpha$ ($f$ is *onto*). We require every $Rs \in \mathcal{R}$ to be a counterpart of some resolution logic. If, in addition, for every finite $X \subseteq L$, clean in every system of $\mathcal{R}$,

(r4) $$C(X) = L \textit{ iff } \alpha[r/X \overset{f(r)}{\Rightarrow} \square] \textit{ is true},$$

then $\mathfrak{R}$ is called an AND-OR *resolution circuit* for $\mathcal{P}$. As in our previous example, the expression $\alpha[r/X \overset{f(r)}{\Rightarrow} \square]$ denotes the result of simultaneous replacement of all occurrences of every variable $r$ of $\alpha$ by the expression $X \overset{f(r)}{\Rightarrow} \square$. We call $\alpha$ the *type* of $\mathfrak{R}$. Finally, we say that an AND-OR resolution circuit $\mathfrak{R}$ for $\mathcal{P}$ is *minimal* if for every AND-OR resolution circuit $\mathfrak{R}' = \langle \mathcal{R}', \alpha', f' \rangle$ for $\mathcal{P}$ there is $Rs \in \mathcal{R}'$ such that the size of $Rs$ is $\geq$ to the size of every resolution system of $\mathfrak{R}$. In other words, $\mathfrak{R}$ consists of the smallest systems possible.

When comparing the definition of a resolution counterpart with the definition of a resolution circuit we may view the condition (r4) as an analog of the refutational completeness condition (r1). The conditions (r2) and (r3), however, are not reflected in the definition of the resolution circuit (see Exercise 6.5).

LEMMA 6.1: *Let $\langle \mathcal{R}, \alpha, f \rangle$ be an* AND-OR *resolution circuit for a resolution logic $\mathcal{P}$ and let $\beta$ be an and-or formula equivalent to $\alpha$. Then, $\langle \mathcal{R}, \beta, f \rangle$ is a resolution circuit for $\mathcal{P}$.*

*Proof:* Let $\mathcal{P}, \langle \mathcal{R}, \alpha, f \rangle$, and $\beta$ be as stated. Without any loss of generality we can assume that $Var(\alpha) = Var(\beta)$. The lemma follows from the fact that for every finite set $X \subseteq L$, the following sentences are equivalent:

$$\alpha[r/X \overset{f(r)}{\Rightarrow} \Box] \textit{ is true;}$$

$$\beta[r/X \overset{f(r)}{\Rightarrow} \Box] \textit{ is true.}$$

■

THEOREM 6.2: *For every* AND-OR *resolution circuit* $\Re$ *on* $\mathcal{L}$ *there exists a resolution logic* $\mathcal{P}$ *such that* $\Re$ *is an* AND-OR *circuit for* $\mathcal{P}$.

*Proof:* Let $\Re = \langle \mathcal{R}, \alpha, f \rangle$ be a resolution circuit on $\mathcal{L}$. By Lemma 6.1 we can assume that $\alpha$ is in conjunctive normal form. Suppose that $\alpha = \beta_0 \wedge \ldots \wedge \beta_n$. For every $r \in Var(\alpha)$, let $M_r$ be the matrix induced by $f(r)$. By Theorem 3.5, $f(r)$ is a counterpart of $\langle \mathcal{L}, C_{M_r} \rangle$. If $\beta = r_0 \vee \ldots \vee r_k$ is a conjunct of $\alpha$, then let $M_\beta$ be the direct product $M_{r_0} \times \ldots \times M_{r_k}$, and let $C_\beta$ denote the consequence operation determined by $M_\beta$. Finally, let $M$ be the ramified product $M_{\beta_0} \otimes \ldots \otimes M_{\beta_n}$. By Proposition 1.12, $C_M(X) = L$ if and only if for every conjunct $r_0 \vee \ldots \vee r_k$ of $\alpha$, there is $j \leq k$, such that $C_{M_{r_j}}(X) = L$. Hence, $\Re$ is a resolution circuit for $\mathcal{P} = \langle \mathcal{L}, C_M \rangle$. Moreover, by Proposition 1.12 and Theorem 3.7, $\mathcal{P}$ is a resolution logic. ■

Theorem 6.2 shows that any AND-OR network of resolution proof systems corresponds to at least one resolution logic. Similarly, we can easily argue that:

THEOREM 6.3: *For every and-or formula* $\alpha$ *there exists a resolution logic* $\mathcal{P}$ *such that a minimal* AND-OR *resolution circuit for* $\mathcal{P}$ *is of type* $\alpha$.

This result is, in a certain sense, the best possible. The notion of a resolution circuit cannot be easily expanded to 'Boolean circuits' (by the addition of NOT-branches) without violating Theorem 6.2; in general, Boolean circuits do not correspond to logical calculi. The proof of Theorem 6.3 can be based on the observation that if $\alpha$ is an *and-or* formula, then $\Re = \langle \{Rs_2\}, \alpha, f \rangle$ is a minimal AND-OR circuit for classical propositional logic $\mathcal{P}_2$, where $Rs_2$ is the two-verifier resolution counterpart of $\mathcal{P}_2$, and for every variable $r$ of $\alpha$, $f(r) = Rs_2$. Of course, we are cheating a little bit at this point by allowing superfluous occurrences of $Rs_2$ in $\Re$; $\Re$ is minimal but not 'optimal'.

The size of a minimal resolution counterpart can be the most serious obstacle to an efficient implementation of the deductive process. The principal motivation for introducing AND-OR circuits is to provide a resolution based theorem proving environment for a number of resolution logics whose minimal resolution counterparts are of sizes too 'devastating' to even attempt an implementational project. Let us look at one example of such a logic.

EXAMPLE 6.4: The logic $\mathcal{P} = \langle \mathcal{L}, C_M \rangle$ we discuss in this example has as its language the language of classical propositional logic. To define the consequence operation $C_M$ we need a sequence $q_0, q_1, \ldots, q_k$ of different prime numbers. Then, for every $i \leq k$, let $M_i^*$ be the $q_i$-valued Post matrix (Post matrices are defined in Example 1.6). Finally, we let $C_M$ to be the consequence operation defined by the matrix

$$M = M_0^* \times M_1^* \times \ldots \times M_k^*.$$

This completes the definition of $\mathcal{P}$. One sad fact about $\mathcal{P}$ is that every minimal resolution counterpart of $\mathcal{P}$ has at least $q = q_0 \cdot q_1 \cdots q_k$ verifiers. (Note that by Theorem 3.7, $\mathcal{P}$ is a resolution logic.) To prove this fact, let $Rs$ be a minimal resolution counterpart of $\mathcal{P}$ and let $v_T$ be any verifier of $Rs$ such that $v_T \in C_M(\emptyset)$ ($v_T$ can be selected by taking any verifier $v$ of $Rs$ and transforming the formula $v \to v$ in $Rs$). We claim that

(a) for every $i, j \leq q$, if $i \neq j$, then $\neg^i v_T$ and $\neg^j v_T$ are not $\Theta^R_{C_M}$-congruent.

(The meaning of '$\neg^k$' is explained in Example 1.6.) For the proof, let $i$ and $j$ be as required, say $j > i$, and let $\gamma(p) = \neg^{(q-j)} p$. Our aim is to show that

(b) $C_M(\gamma(p/\neg^j v_T)) \neq L$,

while

(c) $C_M(\gamma(p/\neg^i v_T)) = L$.

Since $\gamma(p/\neg^j v_T) = \neg^q v_T$, by the choice of $q$, $\neg^q v_T \in C_{M_l^*}(\emptyset)$, for every $l \leq k$. So, by Proposition 1.12(ii), $\neg^q v_T \in C_M(\emptyset) \neq L$. This proves (b). To show (c), let us first note that

(d) for some $l \leq k$, $q_l$ is not a divisor of $q - j + i$.

If (d) were false, then, since $q_0, \ldots, q_k$ are different primes, we would be able to select an integer $s$ such that

$$q - j + i = s \cdot q_0 \cdot q_1 \cdots q_k,$$

which is clearly not possible. Having (d) in our disposal, we prove (c) in the following way. If $C_{M_l^*}(\gamma(p/\neg^i v_T)) \neq L$, where $l$ is an integer satisfying (d), then we would have

$$\neg^{(q-j+i)} v_T \in C_{M_l^*}(\emptyset).$$

So, for some $s, q - j + i = s \cdot q_l$, which contradicts the choice of $l$. This shows that $C_{M_l^*}(\gamma(p/\neg^i v_T)) = L$, from which (c) follows by Proposition 1.12(ii). From (b) and (c) we get (a), which, in turn, implies that $Rs$ has at least $q$ different verifiers.

Now, let $r_0, r_1, \ldots, r_k$ be a list of different propositional variables, and let $\Re = \langle\{R_i : i \leq k\}, \alpha, f\rangle$ be a resolution circuit on $\mathcal{L}$ defined as follows. For every $i \leq k$, $R_i$ is a resolution counterpart of the $q_i$-valued Post logic and $f(r_i) = R_i$. Every $R_i$ can be constructed, for instance, in the way described in the paragraph following the proof of Theorem 4.8 (see also Appendix E). Moreover, $\alpha = r_0 \vee r_1 \vee \ldots \vee r_k$ (cf. Figure 6.3). The fact that $\Re$ is a resolution circuit for $\mathcal{P}$ follows from Proposition 1.12(ii).

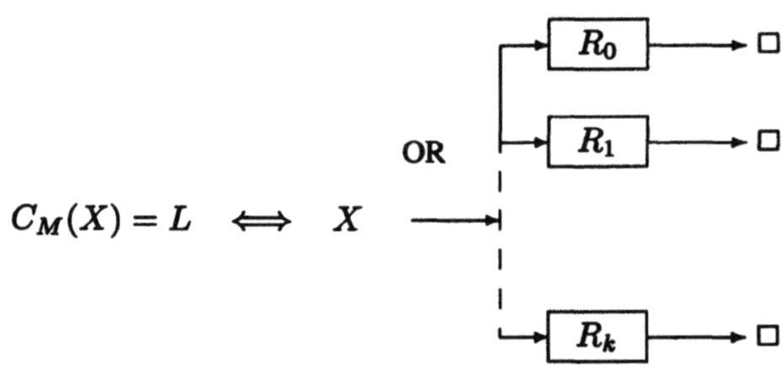


FIG. 6.3: AND-OR resolution circuit for $\mathcal{P}$.

A similar result can be obtained if, instead of Post matrices, we use modified Łukasiewicz matrices in which the negation connective is interpreted in the same way as in Post matrices. ∎

The construction of the logic $\mathcal{P}$ given in the above example proves the following fact.

THEOREM 6.5: *For every sequence $q_0, q_1, \ldots, q_k$ of different primes, $k > 0$, there exists a resolution logic $\mathcal{P}$ such that:*

(i) *every minimal resolution counterpart of $\mathcal{P}$ has at least $q_0 \cdot q_1 \cdots q_k$ verifiers;*

(ii) *there exists a minimal AND-OR resolution circuit $\mathfrak{R}$ for $\mathcal{P}$ such that the size of every resolution proof system in $\mathfrak{R}$ is $\leq max(q_0, q_1, \ldots, q_k)$.*

Now, let us turn to the problem of the effective construction of resolution circuits. We say that an AND-OR resolution circuit $\mathfrak{R} = \langle \mathcal{R}, \alpha, f \rangle$ is *optimal* for a resolution logic $\mathcal{P}$ if $\mathfrak{R}$ is minimal for $\mathcal{P}$ and $\alpha$ has the smallest length among the types of minimal resolution circuits for $\mathcal{P}$. Clearly, every resolution logic has an optimal resolution circuit. In some cases, optimal resolution circuits consist of just one resolution proof system. For example, the optimal AND-OR circuit for classical propositional logic consists of a single two-verifier resolution proof system $Rs_2$ (see Appendix A). As in the case of minimal resolution counterparts, optimal resolution circuits for resolution logics can be effectively constructed. Before we prove this fact, let us introduce the following notation. We say that two resolution circuits on $\mathcal{L}$ are *refutationally equivalent* if both are AND-OR circuits for the same resolution logic.

THEOREM 6.6: *There is an algorithm to determine whether or not two* AND-OR *circuits are refutationally equivalent.*

*Proof:* Let $\mathfrak{R}_i = \langle \mathcal{R}_i, \alpha_i, f_i \rangle, i = 0, 1$, be two resolution circuits on $\mathcal{L}$ and let $M_i$ be the matrix constructed from $\mathfrak{R}_i$ as in the proof of Theorem 6.2. Clearly, $\mathfrak{R}_0$ and $\mathfrak{R}_1$ are refutationally equivalent if and only if the consequence operations $C_{M_i}, i = 0, 1$, have the same finite inconsistent sets. Let $M$ be the direct and $N$ the ramified product of $M_0$ and $M_1$. $\mathfrak{R}_0$ and $\mathfrak{R}_1$ are refutationally equivalent if and only if

(a) $C_M = C_N$.

By Proposition 1.11, the truth-value of (a) can be effectively determined. ∎

In the proof of the next theorem we describe our first algorithm for selecting optimal resolution circuits. We call a resolution circuit $\mathfrak{R}$ for $\mathcal{P}$ *strong*, if every system $Rs$ of $\mathfrak{R}$ is a strong resolution counterpart of the logic it induces.

THEOREM 6.7: *There is an effective method which for every resolution logic $\mathcal{P}$ with a strong resolution counterpart constructs its optimal strong* AND-OR *resolution circuit.*

*Proof:* Let $\mathcal{P} = \langle \mathcal{L}, C \rangle$ be a resolution logic and let $Rs^*$ be a strong resolution counterpart of $\mathcal{P}$ of size $m$. Moreover, let

(a) $N_0, \ldots, N_s$

be a sequence of all non-isomorphic matrices of cardinality $\leq m$. Clearly, if $\Re$ is a minimal strong resolution circuit for $\mathcal{P}$, then isomorphic copies of the matrices induced by the resolution proof systems of $\Re$ are in (a). Let $t = m^{2^m+m-1}$, and let $\Theta$ be the congruence of $\mathcal{L}$ defined as follows: for every $\alpha, \beta \in L$,

$$\alpha \, \Theta \, \beta \text{ iff for every } i \leq s \text{ and every } h \in Hom(\mathcal{L}, N_i), h(\alpha) = h(\beta).$$

In other words, $\alpha$ and $\beta$ are $\Theta$-congruent if and only if they cannot be distinguished in any of the matrices listed in (a). Since all the matrices of (a) are finite, $L^{(t)}/\Theta$ is finite and we can effectively select a finite set $W \subseteq L^{(t)}$ such that every formula of $L^{(t)}$ is $\Theta$-congruent to some formula in $W$. Moreover, we assume that every constant symbol of $\mathcal{L}$ is in $W$. Finally, let us list all resolution circuits constructed from formulas of $W$. Since this list is finite, we can select an optimal circuit among all the circuits that can be constructed from the resolution proof systems on this list.

The correctness of the above algorithm can be demonstrated as follows. Let $\Re = \langle \mathcal{R}, \alpha, f \rangle$ be an optimal strong resolution circuit for $\mathcal{P}$ and let $Rs \in \mathcal{R}$. Clearly, $Rs$ has $\leq m$ verifiers. By Theorem 4.6, we may assume that the verifiers of $Rs$ belong to $L^{(t)}$. Let $N$ be one of the matrices listed in (a) which is an isomorphic copy of the matrix induced by $Rs$. Since $\Theta \subseteq \Theta^R_{C_N}$, by Theorem 2.11, the replacement of any verifier of $Rs$ by a $\Theta$-congruent formula does not change refutational properties of $Rs$. So, we may also assume that the verifiers of $Rs$ are among the formulas of $W$.

Next, let

$$R_0, \ldots, R_l$$

be an enumeration of all different resolution proof systems on $\mathcal{L}$, having $\leq m$ verifiers, that can be constructed using the formulas of $W$. We can eliminate from this list all such systems that are not proper and are not strong counterparts of the logics they induce (see Theorem 3.2 and Exercise 3.6). Let the resulting list be

$$R_0, \ldots, R_k,$$

and let $r_0, \ldots, r_k$ be a list of different propositional variables. There are only finitely many non-equivalent *and-or* formulas built up from the variables in question, and all these formulas can be constructed effectively (see Exercise 6.3). The number of different functions mapping $\{r_0, \ldots, r_k\}$ onto $\{R_0, \ldots, R_k\}$ is also finite. Hence, we can only construct finitely many strong resolution circuits using the formulas of $W$.

Finally, we select an optimal circuit by eliminating those resolution circuits that are not refutationally equivalent to $\langle\{Rs^*\}, r_0, f\rangle$, where $f(r_0) = Rs^*$, and this can be done in the way described in the proof of Theorem 6.6. ∎

By inspecting the proof of Theorem 6.7, we conclude that both occurrences of 'strong' can be dropped from the wording of this theorem, if the upper bound $t$ on the number of variables sufficient to construct an optimal resolution circuit for a given resolution logic is known in advance. Hence:

THEOREM 6.8: *There is an effective method which for every resolution logic $\mathcal{P}$ and every integer $t \geq 0$ constructs a resolution circuit which is optimal among the circuits $\mathfrak{R}$ for $\mathcal{P}$ such that the verifiers of every resolution proof system of $\mathfrak{R}$ are in $L^{(t)}$ (provided that such a circuit exists).*

The situation described in this theorem is analogous to the way minimal resolution counterparts can be found using the minimization algorithm of Theorem 4.5.

Let $\langle\{Rs\}, r, f\rangle$ be a resolution circuit for a logic $\mathcal{P}$. Naturally, one may expect $Rs$ to be a resolution counterpart of $\mathcal{P}$. Our next theorem shows that this is precisely the case.

THEOREM 6.9: *Let $\mathcal{P}$ be a resolution logic on $\mathcal{L}$ and let $\mathfrak{R} = \langle\{Rs\}, r, f\rangle$ be a resolution circuit on $\mathcal{L}$, where $r$ is a propositional variable and $f(r) = Rs$. Then $\mathfrak{R}$ is a resolution circuit for $\mathcal{P}$ iff $Rs$ is a resolution counterpart of $\mathcal{P}$.*

*Proof:* Let $\mathcal{P}$ and $\mathfrak{R}$ be as stated. Clearly, if $Rs$ is a resolution counterpart of $\mathcal{P}$, then $\mathfrak{R}$ is a resolution circuit for $\mathcal{P}$. Conversely, let us assume that $\mathfrak{R}$ is a resolution circuit for $\mathcal{P}$. In this case (r4) and (r1) are equivalent. Since $Rs$ is a resolution counterpart of the logic $\mathcal{P}_{Rs}$ it induces, both $\mathcal{P}$ and $\mathcal{P}_{Rs}$ have the same inconsistent sets. From this fact we get (r2) and (r3) immediately. ∎

In practical applications, we may be forced to work with resolution circuits that are not optimal, and this can happen for a variety of reasons. First, the execution of the algorithm described in Theorem 6.7 (or similar algorithms) to find an optimal circuit may prove to be too costly. Therefore, an optimal resolution circuit for a particular logic may simply not be available to us. Second, the number of processors that are at our disposal may not be sufficiently large to implement a particular optimal resolution circuit. It turns out, however, that non-optimal resolution circuits can be 'locally optimized' when, e.g., additional processors are available to us. Let us look at the resolution circuit $\mathfrak{R}$ pictured in Figure 6.2. Clearly, $\mathfrak{R}$ may not be optimal for $\mathcal{P}$. Let us suppose that, at some point, we learn that the logic induced by $R_2$ has the following resolution circuit

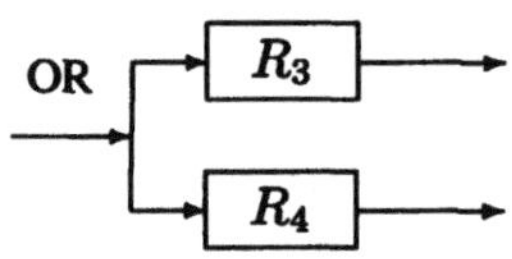


FIG. 6.4: Resolution circuit for the logic induced by $R_2$.

and that both $R_3$ and $R_4$ have fewer verifiers than $R_2$. Let us replace $R_2$ in $\mathfrak{R}$ with the circuit graphed in Figure 6.4 (see Figure 6.5).

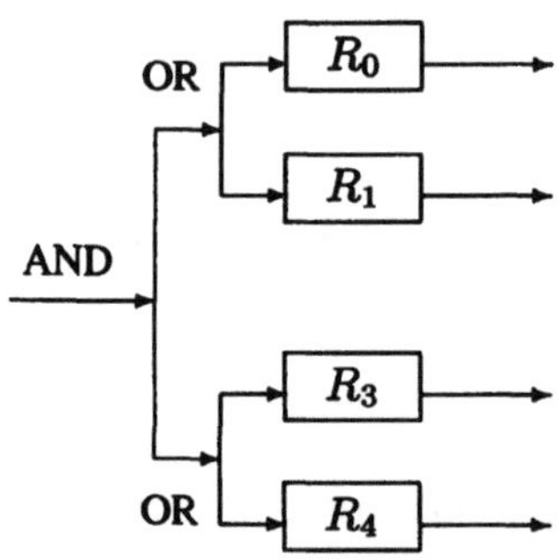


FIG. 6.5: Modified AND-OR resolution circuit depicted in Figure 6.2.

Is the new circuit a resolution circuit for $\mathcal{P}$? The following theorem provides the affirmative answer to this question.

THEOREM 6.10: *Let $\langle \mathcal{R}_0, \alpha, f\rangle$ be a resolution circuit for $\mathcal{P}$, let $Rs \in \mathcal{R}_0$, and let $\langle \mathcal{R}_1, \beta, g\rangle$ be a resolution circuit for the logic induced by Rs. Moreover, let us assume that $Rs \notin \mathcal{R}_1$, that $Var(\alpha) \cap Var(\beta) = \emptyset$, and that $r_i$ is a variable of $\alpha$ such that $f(r_i) = Rs$. Then the circuit $\langle \mathcal{R}, \alpha[r_i/\beta], h\rangle$ is a resolution circuit for $\mathcal{P}$, where:*

(i) $\mathcal{R} = (\mathcal{R}_0 \cup \mathcal{R}_1) - \{Rs\}$;

(ii) *the function h mapping $Var(\alpha[r_i/\beta])$ onto $\mathcal{R}$ is defined as follows:*
*if $r \in Var(\alpha)$, then $h(r) = f(r)$, else $h(r) = g(r)$.*

*Proof:* Let $\mathcal{P} = \langle \mathcal{L}, C\rangle$. By (r4), for every finite $X \subseteq L$, clean in every system of $(\mathcal{R}_0 \cup \mathcal{R}_1) - \{Rs\}$,

(a) $C(X) = L$ *iff* $\alpha[r/X \overset{f(r)}{\Longrightarrow} \square]$ *is true,*

and

(b) $X \overset{f(r_i)}{\Longrightarrow} \square$ *iff* $\beta[q/X \overset{g(q)}{\Longrightarrow} \square]$ *is true.*

From (a) and (b) our theorem follows immediately. ∎

Let us close this section with the following remark. Although AND-OR circuits provide us with an 'optimal' resolution representation of propositional resolution logics, there still exist 'hard' logical calculi whose optimal AND-OR circuits consist of resolution proof systems of sizes far too large for efficient implementation. Since every algorithm satisfying Theorem 6.6 must print all the resolution proof systems of such optimal AND-OR circuits, the practical applicability of such general purpose methods is of limited value unless, as in the case of the generation of minimal resolution counterparts, we restrict our search to specific subclasses of resolution logics.

**EXERCISES**

6.1. Let us call a system $\mathfrak{R} = \langle \mathcal{R}, \alpha, f\rangle$ a *Boolean resolution circuit*, if $\mathcal{R}$ and $f$ are as in the definition of a resolution circuit and $\alpha$ is an arbitrary formula of classical propositional logic constructed from propositional variables and the connectives $\vee, \wedge$, and $\neg$. If $\mathcal{P} = \langle \mathcal{L}, C\rangle$ is a propositional logic such that for every finite $X \subseteq L$, clean in every resolution proof system of $\mathcal{R}$,

$$C(X) = L \text{ iff } \alpha[r/X \overset{f(r)}{\Rightarrow} \Box] \text{ is true,}$$

then $\mathfrak{R}$ is called a *Boolean resolution circuit* for $\mathcal{P}$. Show that there exist Boolean resolution circuits that are circuits for no resolution logic.

6.2. Let $\langle \mathcal{R}, \alpha, f \rangle$ be an AND-OR resolution circuit for $\langle \mathcal{L}, C \rangle$. For every $Rs \in \mathcal{R}$, let $C_{Rs}$ denote the consequence operation determined by the matrix induced by $Rs$. Show that $\bigcap\{\Theta^R_{C_{Rs}} : Rs \in \mathcal{R}\} \subseteq \Theta^R_C$.

6.3. Design an algorithm which for every formula $\alpha$ of classical propositional logic constructs an *and-or* formula equivalent to $\alpha$, if such a formula exists.

6.4. Show that if a resolution circuit $\langle \{Rs\}, \alpha, f \rangle$ is optimal for $\mathcal{P}$, then $Rs$ is a minimal resolution counterpart of $\mathcal{P}$.

6.5. Let $\mathfrak{R} = \langle \{R_i : i \leq k\}, r_0 \wedge \ldots \wedge r_k, f \rangle$ be a resolution circuit for $\langle \mathcal{L}, C \rangle$, where for every $i \leq k$, $f(r_i) = R_i$. Show that for every $i \leq k$, and every set $V$ of verifiers of $R_i = \langle \mathcal{V}_i, \mathcal{F}_i \rangle$, $C(V) = L$ implies $V \in \mathcal{F}_i$.

6.6. Let $Rs_n$ be the resolution counterpart of the $n$-valued Łukasiewicz logic $\mathcal{P}_n$, defined in Section 4.3. Moreover, suppose that $n - 1$ is prime. Show that $\langle \{Rs_n\}, r_0, f \rangle$, where $f(r_0) = Rs_n$, is an optimal resolution circuit for $\mathcal{P}_n$.

# 7

# FIRST-ORDER RESOLUTION PROOF SYSTEMS

In this chapter we extend the theory of propositional resolution proof systems to finitely-valued first-order logics. Throughout this chapter we assume that all $FFO$ logics under consideration have infinitely many ground atomic formulas and that these logics are defined by proper and well-connected matrices.

## 7.1 UNIFICATION

Before we introduce the main notion of this chapter, the notion of a first-order resolution proof system, we shall briefly discuss the notion of a unifier of a set of atomic formulas of an $FFO$ logic. We shall employ this notion in the definition of the resolution rule of a first-order resolution proof system. To avoid a lengthy exposition of facts well-known from the literature on automated reasoning, we shall only provide a list of definitions with rather scarce commentaries. The reader may refer to [58] for a comprehensive exposition of the unification theory.

Let $\mathcal{L}$ be a fixed first-order language, let $X_L$ be the set of individual variables of $\mathcal{L}$, and let $T_L$ be the set of terms of this language. A *term substitution* is any function $\kappa$ mapping $X_L$ into $T_L$ such that $\kappa(x) \neq x$ for only finitely many $x \in X_L$. If $\alpha \in L$ and $\kappa$ is a term substitution, then $\kappa(\alpha)$ denotes the formula obtained from $\alpha$ by simultaneous replacement of every occurrence of every variable $x$ of $\alpha$ by $\kappa(x)$. For example, if $\alpha = R(f(x,y),a)$, where $x$ and $y$ are the only individual variables occurring in $\alpha$, and if $\kappa$ is a term substitution such that $\kappa(x) = f(y,b)$ and $\kappa(y) = x$, then $\kappa(\alpha) = R(f(f(y,b),x),a)$.

We say that a finite set $X = \{\phi_0,\ldots,\phi_l\}$ of atomic formulas is *unifiable* if there is a term substitution $\kappa$, called a *unifier* of $X$, such that $\kappa(\phi_0) = \ldots = \kappa(\phi_l)$. A unifier $\kappa$ of $X$ is called *most general*, if for every other unifier $\eta$ of $X$ there is a term substitution $\varepsilon$ such that

$$\eta = \kappa \circ \varepsilon.$$

The substitution $\kappa \circ \varepsilon$, i.e., the composition of $\kappa$ and $\varepsilon$, is defined by: for every $x \in X_L$,

$$\kappa \circ \varepsilon(x) = \varepsilon(\kappa(x)).$$

PROPOSITION 7.1 (Robinson, cf. [54]): *There is an algorithm which for every finite and non-empty set $X$ of atomic formulas constructs a most general unifier of $X$, if this set is unifiable, and returns* '**not unifiable**', *otherwise.*

Examples of unification algorithms satisfying Proposition 7.1 can be found in [7, 32, 54, 58]. Throughout this chapter, the terms 'substitution' and 'term substitution' are assumed to be synonymous.

Let $\kappa$ be a substitution, let $\alpha \in L$, and let $V$ be a finite set of ground formulas of $\mathcal{L}$. We say that $\kappa$ is *non-creative* for $\alpha$ with respect to $V$ if for every subformula $\phi$ of $\alpha$, $\kappa(\phi) \in V$ implies $\kappa(\phi) = \phi$. In other words, if $\kappa$ is non-creative for $\alpha$ with respect to $V$, then $\kappa$ cannot introduce new occurrences of formulas of $V$ to $\kappa(\alpha)$. We say that a formula $\alpha$ is a *non-creative instance* of $\beta$ with respect to a finite set $V$ of ground formulas, if $\alpha = \kappa(\beta)$, for some non-creative substitution $\kappa$ for $\beta$ with respect to $V$.

EXAMPLE 7.2: Let us assume that among the symbols of $\mathcal{L}$ there is a binary predicate $<$ and two constant symbols 0 and 1. Let $\beta$ be $(0 < 1) \vee (x < y)$ and let $V = \{0 < 1\}$. Any substitution $\kappa_0$ such that $\kappa_0(x) = 0$ and $\kappa_0(y) = 1$ is creative for $\beta$ since $\kappa_0(x < y) = 0 < 1$. In other words, $\kappa_0$ 'creates' a new occurrence of $0 < 1$ in $\kappa_0(\beta)$ in addition to the one already in $\beta$. On the other hand, any substitution $\kappa_1$ such that $\kappa_1(x) = 1$ and $\kappa_1(y) = 0$ is non-creative. Hence, $(0 < 1) \vee (1 < 0)$ is a non-creative instance of $\beta$. ∎

## 7.2 RESOLUTION COUNTERPARTS OF FFO LOGICS

The notion of a first-order resolution proof system we are about to introduce generalizes the concept of a resolution proof system for a propositional logic introduced in Chapter 2 and the (basic) resolution proof system for classical predicate logic proposed in [30] (see also [39]). The non-clausal resolution proof system for quantifier-free first-order logic studied by Manna and Waldinger in [30] is based on two types of inference rules:

- the transformation rules, i.e., expressions of the form $\phi \Rightarrow v$, which allow to simplify deduced formulas by replacing occurrences of $\phi$ with inferentially equivalent but syntactically simpler formula $v$;
- the resolution rule, i.e., the set of all sequents of the form

$$\frac{\alpha(\phi), \beta(\psi)}{\kappa(\alpha)(\kappa(\phi)/F) \vee \kappa(\beta)(\kappa(\psi)/T)}.$$

In this sequent, $\alpha$ and $\beta$ are arbitrary formulas that do not share variables (are standardized apart); $\phi$ and $\psi$ are subformulas of $\alpha$ and $\beta$, respectively; $\kappa$ is a most general unifier of $\{\phi, \psi\}$ (hence $\kappa(\phi) = \kappa(\psi)$). The conclusion is obtained by replacing all occurrences of $\kappa(\phi)$ in $\kappa(\alpha)$ by $F$, replacing all occurrences of $\kappa(\psi)$ in $\kappa(\beta)$ by $T$, and taking the disjunction of the results; $T$ and $F$ are formulas defining *truth* and *falsehood*, respectively.

This resolution proof system operates on a finite set of quantifier-free formulas called the *deduced set*. Similarly to the propositional case, the deductive process can be viewed as a sequence of applications of the inference rules to formulas of the deduced set. Each such application adds new formulas to the deduced set without altering its consistency status. This deductive process terminates successfully as soon as $F$ is added to the deduced set.

Having developed the notion of a propositional resolution proof system, we can model the notion of a resolution counterpart of an $FFO$ logic on the example just discussed. Let $\mathcal{L}$ be a first-order language with the disjunction connective denoted by $\vee$. A *resolution proof system* on $\mathcal{L}$ is a resolution algebra of the form $Rs = \langle \mathcal{V}, \mathcal{F} \rangle$, where $\mathcal{V}$ is a finite algebra of ground formulas of $\mathcal{L}$, whose operations correspond to the connectives of $\mathcal{L}$, and $\mathcal{F}$ is a nonempty family of subsets of the base set of $\mathcal{V}$. As in the propositional case, we call the elements of the algebra $\mathcal{V}$ *verifiers* and, as it was done in the preceding chapters, we reserve the symbol '$Ver_{Rs}$' to denote the set of all verifiers of $Rs$.

With every resolution proof system $Rs$ we associate the set of *transformation rules*, defined as in Section 2.2, and the *resolution rule* – an inference rule consisting of all sequents of the form

$$\frac{\alpha_0(\phi_0), \ldots, \alpha_n(\phi_n)}{\kappa(\alpha_0)(\kappa(\phi_0)/v_0) \vee \ldots \vee \kappa(\alpha_n)(\kappa(\phi_n)/v_n)}.$$

In this sequent, $\alpha_0, \ldots, \alpha_n$ are standardized apart, $\phi_i$ is an atomic subformula of $\alpha_i$, $\kappa$ is a most general unifier of $\{\phi_0, \ldots, \phi_n\}$, and $\kappa(\phi_0)$ is not a subformula of a verifier of $Rs$. The conclusion (called the *resolvent* of $\alpha_0, \ldots, \alpha_n$) is obtained by replacing all occurrences of $\kappa(\phi_i)$ in $\kappa(\alpha_i)$ with the $i$-th verifier $v_i$, and, then, by taking the disjunction of the results.

EXAMPLE 7.3: Let $\mathcal{L}$ be a first-order language having the following symbols:

$$P_L = \{P, Q, R\},$$
$$F_L = \{b, f\},$$
$$Con_L = \{\neg, \vee, \wedge, \rightarrow\},$$

where $P$, $Q$, $R$, and $f$ are unary and $b$ is the only constant symbol of $\mathcal{L}$. The connectives of $\mathcal{L}$ have their usual meanings. Let the resolution proof system $Rs_3^f = \langle \mathcal{V}, \mathcal{F} \rangle$ on $\mathcal{L}$ be defined as follows. The verifiers of $Rs_3^f$ are the formulas:

$$v_5 = Q(b) \rightarrow Q(b), \quad v_4 = Q(b) \vee \neg Q(b),$$
$$v_3 = \neg(v_4 \rightarrow \neg v_4), \quad v_2 = \neg v_3, \quad v_1 = \neg v_4, \quad v_0 = \neg v_5.$$

The transformation rules of $Rs_3^f$ and the family $\mathcal{F}$ are defined in the same way as for the propositional resolution proof system $Rs_3$ described in Appendix B. ■

Let $Rs$ be a resolution proof system on $\mathcal{L}$ and let $X \subseteq L$. As in the propositional case, we shall write $X \overset{Rs}{\Rightarrow} \Box$ to denote the existence of a refutation of $X$ in $Rs$, i.e., the existence of a sequence $\alpha_0, \ldots, \alpha_m$ of formulas such that each $\alpha_i$ either is in $X$ or is obtained from some formulas in the sequence earlier than $\alpha_i$ with the help of an inference rule of $Rs$. Moreover, some subset $V$ of $\{\alpha_0, \ldots, \alpha_m\}$ is a member of $\mathcal{F}$.

For reasons explained in Section 3.4, the deductive process in propositional resolution proof systems is restricted to clean sets only. As a result, we can view verifiers of such systems as special constants: we can neither resolve upon variables occurring in verifiers nor use verifiers to represent a reasoning task. To treat verifiers of first-order resolution proof systems in a similar way, we have already set-up the definition of the resolution rule in such a way that the resolution upon subformulas of verifiers is prohibited. Furthermore, we should require that during the deductive process verifiers cannot be 'introduced' to formulas through unification. In other words, unifiers which are used in the formation of resolvents should be non-creative with respect to the set of verifiers of a given resolution proof system. This may look like a complicated requirement that introduces the necessity of an extra test to be performed on formulas which we want to resolve. However, it can be easily satisfied through an appropriate definition of a clean set.

> *A finite set $X \subseteq L$ is said to be clean in a first-order resolution proof system $Rs$ if the formulas of $X$ do not share constant symbols with the verifiers of $Rs$.*

There is, however, a price for this simple solution; for some logics, the class of clean sets may not be sufficiently large. For instance, if $\mathcal{L}$ has just one constant symbol, as in Example 7.3, then clean sets consist of formulas without constant symbols. Before we propose a solution to this problem, let us recall that in the propositional case, the problem of non-clean sets is solved by renaming propositional variables occurring in the formulas of a non-clean set $X$, so that the resulting set $X'$ is clean. Then, as $X'$ inherits the consistency status from $X$, we simply try to refute $X'$.

In the first-order case, instead of manipulating formulas of a non-clean set, we can rename all the constant symbols occurring in verifiers, selecting for this purpose new constant symbols that do not occur in the formulas of $\mathcal{L}$ (see also Exercise 7.5). In that way all the finite subsets of $L$ are clean and, hence, can be subjected to a refutation process.

The introduction of new symbols into an object language is not an unusual procedure in logic. The elimination of quantifiers by the method of skolemization may serve as one of the many examples. Moreover, as the next theorem shows, the addition of new constant symbols does not modify the class of inconsistent sets of $L$.

THEOREM 7.4: *Let $\langle \mathcal{L}, C \rangle$ be an $M$-valued $FFO$ logic and let $\mathcal{L}_b$ be the language obtained from $\mathcal{L}$ by the addition of a new constant symbol $b$ that does not occur in formulas of $\mathcal{L}$. Moreover, let $C_b$ denote the consequence operation on $\mathcal{L}_b$ defined by $M$. Then, for every $X \subseteq L, C(X) = L$ iff $C_b(X) = L_b$.*

*Proof:* Let $\mathcal{L}, \mathcal{L}_b, C$, and $C_b$ be as stated, let $M = \langle \mathcal{A}, \mathcal{D} \rangle$, and let $X \subseteq L$ be $C$-inconsistent. To show that $C_b(X) = L_b$, let $d \in \mathcal{D}$ and let $\langle U, \pi_b \rangle$ be an $\langle \mathcal{A}, d \rangle$-frame. Let $\langle U, \pi \rangle$ be the $\langle \mathcal{A}, d \rangle$-frame obtained from $\langle U, \pi_b \rangle$ by restricting $\pi_b$ to the symbols of $\mathcal{L}$. Since $C(X) = L$, there is a valuation $v$ of $T_L \cup L$ into $\langle U, \pi \rangle$ such that $v(X) \not\subseteq d$ (cf. Exercise 1.8). Let $v_b$ be the valuation of $T_{L_b} \cup L_b$ into $\langle U, \pi_b \rangle$ such that for every $x \in X_L, v_b(x) = v(x)$. A simple proof by induction on the complexity of a formula shows that for every $\alpha \in L, v_b(\alpha) = v(\alpha)$, and, hence, $v_b(X) \not\subseteq d$. This shows that $C_b(X) = L_b$. In a similar way we show that $C_b(X) = L_b$ implies $C(X) = L$. ∎

EXAMPLE 7.5: Let us return to the definition of the resolution proof system $Rs_3^f$ given in Example 7.3. Let $c$ be a new constant symbol not in $\mathcal{L}$, and let us replace the constant $b$ in every verifier of $Rs_3^f$ by $c$, i.e., the verifiers of $Rs_3^f$ are now:

$$v_5 = Q(c) \rightarrow Q(c), \quad v_4 = Q(c) \vee \neg Q(c),$$
$$v_3 = \neg(v_4 \rightarrow \neg v_4), \quad v_2 = \neg v_3, \quad v_1 = \neg v_4, \quad v_0 = \neg v_5.$$

Formally, after this replacement, $Rs_3^f$ has become a resolution proof system on $\mathcal{L}_c$ and not on $\mathcal{L}$. However, every set $X \subseteq L$ is now clean in $Rs_3^f$. Moreover, by Theorem 7.4, we can consider $\mathcal{L}$ and $\mathcal{L}_c$ identical as long as the constant $c$ is used only in the construction of verifiers. ∎

Let $Rs$ be a resolution proof system on $\mathcal{L}$. We call $Rs$ a *resolution counterpart* of an $FFO$ logic $\langle \mathcal{L}, C \rangle$, if the conditions (r1)–(r3), defined in Section 2.4, are satisfied (in (r2), the relation $\Theta_C^R$ is defined as in the propositional case, see Exercise 7.2 for details).

As in the case of propositional resolution proof systems, resolution counterparts of $FFO$ logics can be explicitly used only to determine the consistency status of the deduced set. In Section 1.2 we argued that the existence of an effective (and preferably efficient) method for reducing the problem of entailment to inconsistency, is one of the fundamental principles on which refutational theorem proving techniques are based. In classical predicate logic this reduction proceeds by first forming the set $X \cup \{\neg\alpha\}$, and, then by eliminating quantifiers from the formulas of this set *via* conversion into prenex normal form and skolemization. The resulting set of quantifier-free formulas is inconsistent if and only if $\alpha$ is a consequence of $X$.

For other first-order logics, the existence of an effective reducibility method based on quantifier elimination depends, among other things, on the type and logical properties of quantifiers as well as on the properties of connectives. For some calculi, such as variants of first-order Post logics studied in [46] or the three-valued logic discussed in [56], quantifier elimination is readily available, while for other logics (e.g. some modal logics) it is not.

While the reduction of entailment to inconsistency through quantifier elimination is one of the most commonly used techniques, there are other directions that might be successfully explored. In [1] it is shown that for some modal logics the process of reducibility may require no, or only some form of quantifier elimination, and that the resolution principle can be defined in terms of quantifier-free as well as quantified formulas.

## 7.3 EXISTENCE OF RESOLUTION COUNTERPARTS

The aim of this section is to provide an effective construction of resolution counterparts of $FFO$ logics. The key to our algorithm is the observation that lifting functions provide connections between the resolution counterparts of a given $FFO$ logic $\mathcal{P}$ and the resolution counterparts of the logic associated with $\mathcal{P}$. The details are as follows.

Let $\mathcal{P} = \langle \mathcal{L}, C \rangle$ be an arbitrary but fixed disjunctive $M$-valued $FFO$ logic, where $M = \langle \mathcal{A}, \mathcal{D} \rangle$ is a proper and well-connected matrix. Following the convention introduced in Section 1.7, by $\mathcal{P}_p = \langle \mathcal{L}_p, C_M \rangle$ we denote the propositional logic associated with $\mathcal{P}$. Let $t$ be a lifting function for $\mathcal{P}$, i.e., a one-one function mapping the set of all atomic formulas of $\mathcal{P}$ onto the set of all propositional variables of $\mathcal{L}_p$. Clearly, if $Rs_p$ is a resolution proof system on $\mathcal{L}_p$, then $Rs$, obtained from $Rs_p$ by replacing every variable $p$ occurring in the verifiers of $Rs_p$ by $t^{-1}(p)$, is a resolution proof system on $\mathcal{L}$. We assume here that for every variable $p$ occurring in the verifiers of $Rs_p$, $t^{-1}(p)$ is a ground atomic formula of $\mathcal{L}$ (if necessary, $p$ can be renamed to satisfy this requirement).

Let $Rs_p$ be an arbitrary resolution counterpart of $\mathcal{P}_p$ and let $Rs$ be the resolution proof system on $\mathcal{L}$ obtained from $Rs_p$ in the way described above. We assume that $v_0^p, \ldots, v_n^p$ is a fixed enumeration of the verifiers of $Rs_p$, while $v_0, \ldots, v_n$ is the enumeration of the verifiers of $Rs$ satisfying $t(v_i) = v_i^p$. It turns out that $Rs$ is a resolution counterpart of $\mathcal{P}$. Our proof of this fact uses Herbrand's Theorem (cf. Theorem 1.16) and the following lemma.

LEMMA 7.6 Lifting Lemma (O'Hearn, Stachniak): *Suppose that $\alpha_0(\phi), \ldots, \alpha_n(\phi)$ are ground instances of $\beta_0, \ldots, \beta_n \in L$, respectively, that these instances are non-creative with respect to $Ver_{Rs}$, and that $\phi$ is an atomic formula which does not occur in the verifiers of Rs. Then there is a resolvent $\gamma$ of $\beta_0, \ldots, \beta_n$ such that $\alpha_0(\phi/v_0) \vee \ldots \vee \alpha_n(\phi/v_n)$ is a non-creative instance of $\gamma$ with respect to $Ver_{Rs}$.*

*Proof*: Let $\lambda_i, i \leq n$, be substitutions which standardize variables of $\beta_0, \ldots, \beta_n$ apart and let $\theta$ be a substitution such that $\theta(\lambda_i(\beta_i)) = \alpha_i$. Moreover, let $W$ be the set of all atomic subformulas of all $\lambda_i(\beta_i)$ which are transformed by $\theta$ into $\phi$, let $\kappa$ be a most general unifier of $W$, and, finally, let $\kappa(W) = \{\phi^*\}$. Since $\theta$ is also a unifier of $W$, $\theta = \kappa \circ \varepsilon$, for some substitution $\varepsilon$. Hence, for every $i \leq n$, $\varepsilon(\kappa(\lambda_i(\beta_i))(\phi^*/v_i)) = \kappa \circ \varepsilon(\lambda_i(\beta_i))(\phi/v_i) = \alpha_i(\phi/v_i)$. This shows that $\alpha_0(\phi/v_0) \vee \ldots \vee \alpha_n(\phi/v_n) = \varepsilon(\kappa(\lambda_0(\beta_0))(\phi^*/v_0) \vee \ldots \vee \kappa(\lambda_n(\beta_n))(\phi^*/v_n))$.

To conclude the proof of this lemma we show that $\varepsilon$ is non-creative with respect to $Ver_{Rs}$. To this end, let $\psi$ be a subformula of $\kappa(\lambda_i(\beta_i))(\phi^*/v_i)$ such that $\varepsilon(\psi) \in Ver_{Rs}$. Let $\psi^*$ be a subformula of $\beta_i$ such that $\psi = \kappa(\lambda_i(\psi^*))$. So,

$\varepsilon(\psi) = \theta(\lambda_i(\psi^*))$ and, since $\lambda_i \circ \theta$ is non-creative for $\beta_i, \varepsilon(\psi) = \psi^* \in Ver_{Rs}$. Hence, $\varepsilon(\psi) = \kappa(\lambda_i(\psi^*)) = \psi$, as required. ∎

THEOREM 7.7 (O'Hearn, Stachniak): *Rs is a resolution counterpart of $\mathcal{P}$.*

*Proof:* Let $Rs_p = \langle \mathcal{V}_p, \mathcal{F}_p \rangle$ and $Rs = \langle \mathcal{V}, \mathcal{F} \rangle$. The proof of (r3) requires only Herbrand's Theorem. Indeed, if $V \subseteq Ver_{Rs}$, then $V \in \mathcal{F}$ *iff* $t(V) \in \mathcal{F}_p$ *iff* $C_M(t(V)) = C_M(t(H(V))) = L_p$ *iff* $C(V) = L$.

To prove (r2), let $f(v_0, \ldots, v_k) \Rightarrow v$ be a transformation rule of $Rs$ and let $X \cup \{\gamma(\phi)\} \subseteq L$. If $C(X, \gamma(\phi/f(v_0, \ldots, v_k))) = L$, then, by Herbrand's Theorem, $C_M(t(H(X \cup \{\gamma(\phi/f(v_0, \ldots, v_k))\}))) = L_p$. Every formula from $t(H(\{\gamma(\phi/f(v_0, \ldots, v_k))\}))$ is of the form $\alpha(t(\phi)/f(t(v_0), \ldots, t(v_k)))$ and, using the transformation rule $f(t(v_0), \ldots, t(v_k)) \Rightarrow t(v)$, is reducible to $\alpha(t(\phi)/t(v)) \in t(H(\{\gamma(\phi/v)\}))$. So, $C_M(t(H(X \cup \{\gamma(\phi/v)\}))) = L_p$ and, by Herbrand's Theorem, $C(X, \gamma(\phi/v)) = L$. In the same way we prove that $C(X, \gamma(\phi/v)) = L$ implies $C(X, \gamma(\phi/f(v_0, \ldots, v_k))) = L$, concluding the proof of (r2).

The proof of (r1) is more complex. Let $A$ be the set of all constant symbols of $\mathcal{L}$ that do not occur in the verifiers of $Rs$. Let $X$ be a finite $C$-inconsistent set of formulas clean in $Rs$. By Herbrand's Theorem there is a finite $C_M$-inconsistent subset $Y$ of $t(H^A(X))$ clean in $Rs_p$. Let

$$(T_p) \qquad \gamma_0, \ldots, \gamma_s$$

be a refutation of $Y$ in $Rs_p$ and for every $i < s$, let $\alpha_i \in L$ be such that $t(\alpha_i) = \gamma_i$. We claim that

$$(T_H) \qquad \alpha_0, \ldots, \alpha_s,$$

is a refutation of $H^A(X)$ in $Rs$. Clearly, if $\gamma_i \in Y$, then $\alpha_i = t^{-1}(\gamma_i) \in H^A(X)$. If

$$\gamma_i = \gamma_{i_0}(p/v_0^p) \vee \ldots \vee \gamma_{i_n}(p/v_n^p)$$

is the resolvent of $\gamma_{i_0}, \ldots, \gamma_{i_n}$, then $t^{-1}(\gamma_{i_j}(p/v_j^p)) = \alpha_{i_j}(t^{-1}(p)/v_j)$, for every $j \leq n$. Hence, $\alpha_i = t^{-1}(\gamma_i)$ is the resolvent of $\alpha_{i_0}, \ldots, \alpha_{i_n}$. Now, if $\gamma_j(p/f(v_0^p, \ldots, v_k^p))$ has been transformed into $\gamma_i(p/v^p)$ using $f(v_0^p, \ldots, v_k^p) \Rightarrow v^p$ of $Rs_p$, then $t^{-1}(\gamma_j(p/f(v_0^p, \ldots, v_k^p))) = \alpha_j(t^{-1}(p)/f(v_0, \ldots, v_k))$ and, using the transformation rule $f(v_0, \ldots, v_k) \Rightarrow v$ of $Rs$, we obtain $\alpha_i(t^{-1}(p)/v) = t^{-1}(\gamma_i(p/v^p))$. To conclude the proof that $(T_H)$ is a refutation of $H^A(X)$ in $Rs$, let us note that some of the formulas $\gamma_{i_0}, \ldots, \gamma_{i_k}$ of $(T_p)$ form a $C_M$-inconsistent subset of verifiers of $Rs_p$. Hence, $\{\alpha_{i_0}, \ldots, \alpha_{i_k}\} \in \mathcal{F}$.

Our next step is to simulate the refutation $(T_H)$ to derive a refutation

$$(T) \qquad \beta_0, \ldots, \beta_s$$

of $X$ in $Rs$ such that

(a) for every $i \leq s, \alpha_i$ is a non-creative instance of $\beta_i$ with respect to $Ver_{Rs}$.

We shall refer to the non-creativity condition (a) when the role of the transformation rules of $Rs$ in the construction of $(T)$ is discussed. If $\alpha_i \in H^A(X)$, then let $\beta_i$ be any formula of $X$ satisfying (a). The non-creativity is guaranteed by the fact that the constants of $A$ do not occur in the verifiers of $Rs$. Next, if

$$\alpha_i = \alpha_{i_0}(\phi/v_0) \vee \ldots \vee \alpha_{i_n}(\phi/v_n)$$

and if $\beta_{i_0}, \ldots, \beta_{i_n}$ have been selected in accordance with (a), then, by Lifting Lemma, $\alpha_i$ is a non-creative instance of a resolvent $\beta_i$ of $\beta_{i_0}, \ldots, \beta_{i_n}$. Finally, if for some $j < i, \alpha_i$ is obtained from $\alpha_j$ by the application of a transformation rule $f(v_0, \ldots, v_k) \Rightarrow v$, then, by (a), $\beta_j$ has an occurrence of $f(v_0, \ldots, v_k)$. So, let $\beta_i$ be the result of the application of the transformation rule in question to $\beta_j$. Clearly, every verifier of $(T_H)$ occurs in $(T)$. Hence, $(T)$ is a refutation of $X$ in $Rs$. This completes the first half of the proof of (r1).

For the second half, let $c_0, \ldots, c_n$ be new symbols which do not occur in formulas of $\mathcal{L}$ and let $\mathcal{L}(c_0, \ldots, c_n)$ be the extension of $\mathcal{L}$ obtained by the addition of $c_0, \ldots, c_n$ as new logical constants. Next, let $N$ be the matrix for $\mathcal{L}_p$ obtained from the matrix $M_p$ induced by $Rs_p$, by replacing every verifier $v_i^p$ with the corresponding constant $c_i$. Clearly, $N$ and $M_p$ are isomorphic and, by Theorem 3.5, the logics defined by $M$ and $N$ have the same inconsistent sets. By Corollary 1.17, the $FFO$ logic defined by $N$ has the same inconsistent sets as $\mathcal{P}$.

There is one more matrix we shall require for the proof. The matrix $N^* = \langle \mathcal{C}, \mathcal{E} \rangle$ is obtained from $N$ by the addition of $\underline{c}_0, \ldots, \underline{c}_n$ as 0-argument operations (i.e., constants). Note, that $N^*$ is not a matrix for $\mathcal{L}_p$ but for $\mathcal{L}_p$ extended by the addition of the logical constants $c_0, \ldots, c_n$. If by $C^*$ we denote the consequence operation on $\mathcal{L}(c_0, \ldots, c_n)$ defined by $N^*$, then, for every $Y \subseteq L$,

(b) $Y$ is $C$-inconsistent *iff* $Y$ is $C^*$-inconsistent.

Next, let $Rs^* = \langle \mathcal{V}^*, \mathcal{F}^* \rangle$ be the resolution proof system on $\mathcal{L}(c_0, \ldots, c_n)$ obtained from $Rs$ by renaming every verifier $v_i$ of $Rs$ as $c_i$ and by adding the constant functions $\underline{c}_0, \ldots, \underline{c}_n$. Clearly, for every $Y \subseteq L$,

(c) $Y \overset{Rs}{\Rrightarrow} \Box$ *iff* $Y \overset{Rs^*}{\Rrightarrow} \Box$.

We claim that, in addition,

(d) if $Y$ is $C^*$-consistent, then $Y$ cannot be refuted in $Rs^*$.

To prove the claim, let us note that every transformation rule of $Rs^*$ is sound with respect to $C^*$. Indeed, every such rule corresponds to an equality true in the algebra of verifiers of $Rs^*$. Moreover, if $V^*$ is a $C^*$-consistent set of verifiers of $Rs^*$, then $V^* \notin \mathcal{F}^*$. Hence, to prove (d) it suffices to demonstrate the soundness of the resolution rule of $Rs^*$. To this end, suppose that for some $d \in \mathcal{E}$, there exists a $\langle \mathcal{C}, d \rangle$-valued model $\mathcal{U}$ of $Z = Y \cup \{\alpha_0(\phi_0), \ldots, \alpha_n(\phi_n)\}$. By Exercise 7.3, we can assume that $\phi = \phi_0 = \ldots = \phi_n$ and that the variables of $\alpha_0, \ldots, \alpha_n$ are standardized apart. Let $h$ be an arbitrary valuation into $\mathcal{U}$. Clearly, $h(Z) \subseteq d$. We claim that

(e) $h(\alpha_0(\phi/c_0) \vee \ldots \vee \alpha_n(\phi/c_n)) \in d$.

Indeed, for some $l \leq n, h(\alpha_l(\phi/c_l)) = h(\alpha_l(\phi)) \in d$, and (e) follows from the well-connectivity of $N^*$. So, $Z$ expanded by the addition of the resolvent of $\alpha_0(\phi), \ldots, \alpha_n(\phi)$ is $C^*$-consistent and, hence, the proof of (d) has been concluded. Finally, the proof that $C(X) \neq L$ implies $X \overset{Rs}{\not\Rrightarrow} \Box$ follows from (b), (c), and (d). ∎

By Theorem 7.7, resolution counterparts of $FFO$ logics can be effectively constructed. Indeed, if $\mathcal{P}$ is an $M$-valued $FFO$ logic, then, first, we construct a resolution counterpart $Rs_p$ of $\mathcal{P}_p$ (cf. the proof of Theorem 3.7). By Theorem 7.7, $Rs_p$ can be transformed into a resolution counterpart $Rs$ of $\mathcal{P}$ when the variables occurring in the verifiers of $Rs_p$ are replaced with ground atomic formulas of $\mathcal{P}$. Finally, to sustain the expressiveness of $\mathcal{P}$, we may replace the constant symbols occurring in the verifiers of $Rs$ by new constants that do not occur in formulas of $\mathcal{P}$.

EXAMPLE 7.8: Let $\mathcal{L}_c$ and the resolution proof system $Rs_3^f$ be as in Example 7.5. By $\mathcal{P}_3^f$ let us denote the $FFO$ logic on $\mathcal{L}_c$ defined by the 3-valued Łukasiewicz matrix. It can be easily demonstrated that $Rs_3^f$ is a resolution counterpart of $\mathcal{P}_3^f$. Indeed, the propositional logic associated with $\mathcal{P}_3^f$ is the 3-valued Łukasiewicz logic $\mathcal{P}_3$. Now, if $t$ is a lifting function for $\mathcal{P}_3^f$ such that $t(Q(c)) = p_0$, then $Rs_3^f$ is the result of replacing all occurrences of $p_0$ in every verifier of $Rs_3$ by $Q(c)$. (See Appendix B for the definition of $Rs_3$). Hence, by Theorem 7.7, $Rs_3^f$ is a resolution counterpart of $\mathcal{P}_3^f$. ■

THEOREM 7.9: *If $Rs$ is a resolution counterpart of an $M$-valued $FFO$ logic $\mathcal{P}$, and if $t$ is a lifting function for $\mathcal{P}$, then the resolution proof system $t(Rs)$ on $\mathcal{L}_p$, obtained from $Rs$ by replacing every verifier $v$ of $Rs$ with $t(v)$, is a resolution counterpart of the logic $\mathcal{P}_p$ associated with $\mathcal{P}$.*

*Proof:* Let $\mathcal{P}, \mathcal{P}_p, t$, and $Rs$ be as stated, and let $X \subseteq L_p$ be finite and clean in $t(Rs)$. By Proposition 1.4, we can assume that $t^{-1}(X)$ consists of ground formulas only and that this set is clean in $Rs$. If $C_M(X) = L_p$, then, by Herbrand's Theorem, $C(t^{-1}(X)) = L$. Let $\alpha_0, \ldots, \alpha_s$ be a refutation of $t^{-1}(X)$ in $Rs$. Since all the formulas occurring in this refutation are ground, $t(\alpha_0), \ldots, t(\alpha_s)$ is a refutation of $X$ in $t(Rs)$. On the other hand, if $X$ is refutable in $t(Rs)$, then, as in the proof of Theorem 7.7, we can construct a refutation of $t^{-1}(X)$ in $Rs$. So, $C(t^{-1}(X)) = L$, and, by Herbrand's Theorem, $C_M(X) = L_p$. This shows that $t(Rs)$ and $\mathcal{P}_p$ satisfy (r1).

To prove (r2), let $f(t(v_0), \ldots, t(v_k)) \Rightarrow t(v)$ be a transformation rule of $t(Rs)$ and let $X \cup \{\gamma(q)\} \subseteq L_p$. If $C_M(X, \gamma(q/f(t(v_0), \ldots, t(v_k)))) = L_p$, then we can assume that $t^{-1}(X \cup \{\gamma(q/f(t(v_0), \ldots, t(v_k)))\})$ consists of ground formulas only (see Proposition 1.4). By Herbrand's Theorem, $C(t^{-1}(X), t^{-1}(\gamma)(t^{-1}(q)/f(v_0, \ldots, v_k))) = L$, from which we get $C(t^{-1}(X), t^{-1}(\gamma)(t^{-1}(q)/v)) = L$, since $f(v_0, \ldots, v_k)$ transforms into $v$ in $Rs$. Using Herbrand's Theorem one more time we conclude that $C_M(X, \gamma(q/t(v))) = L_p$. In the same way we show that

$$C_M(X, \gamma(q/t(v))) = L_p \text{ implies } C_M(X, \gamma(q/f(t(v_0), \ldots, t(v_k)))) = L_p.$$

Finally, the proof of (r3) follows easily from Herbrand's Theorem and is left to the reader. ■

COROLLARY 7.10: *Let $\mathcal{P}, \mathcal{P}_p$, and $t$ be as in Theorem 7.9. If $Rs_p$ is a minimal resolution counterpart of $\mathcal{P}_p$, then $Rs$, obtained from $Rs_p$ by replacing every verifier $v^p$ of $Rs_p$ with $t^{-1}(v^p)$, is a minimal resolution counterpart of $\mathcal{P}$.*

*Proof:* Follows immediately from Theorems 7.7 and 7.9. ∎

By this corollary, to find a minimal resolution counterpart of an $FFO$ logic $\mathcal{P}$ it suffices to find a minimal resolution counterpart of the propositional logic associated with $\mathcal{P}$.

Let us close this section with the following remark. The resolution framework presented in this chapter can also be applied to non-$FFO$ logics. The definition of a resolution counterpart requires no changes to work in the context of other first-order and not just $FFO$ logics. Clearly, the conditions (r1)–(r3) do not require a logic under consideration to be $FFO$.

## 7.4 THEOREM PROVING STRATEGIES

Let $\mathcal{P} = \langle \mathcal{L}, C \rangle$ be an arbitrary $M$-valued $FFO$ logic, fixed throughout this section, and let $\mathcal{P}_p$ be the propositional logic associated with $\mathcal{P}$. Theorem proving strategies discussed in Chapter 5 can be easily adopted to resolution counterparts of $FFO$ logics.

The definition of the set of support strategy does not require any significant modifications. Let $Rs = \langle \mathcal{V}, \mathcal{F} \rangle$ be a resolution counterpart of $\mathcal{P}$.

THEOREM 7.11 (O'Hearn, Stachniak): *Let $X \subseteq L$ be a finite inconsistent set of formulas clean in $Rs$. If $Y$ is a set of support of $X$, then there is an $(X,Y)$-supported refutation of $X$ in $Rs$.*

*Proof:* Let $t$ be a lifting function for $\mathcal{P}$ and let $A$ be the set of all constant symbols of $\mathcal{L}$ that do not occur in the verifiers of $Rs$. Let $Y$ be a set of support of a finite, clean in $Rs$, and inconsistent set $X \subseteq L$. By Herbrand's Theorem:

(a) $C(H^A(X)) = L$ and $C(H^A(X - Y)) \neq L$;

(b) $Z = H^A(X) - H^A(X - Y)$ is a set of support of $H^A(X)$;

(c) there is a finite subset $X_p$ of $t(H^A(X))$ such that $C_M(X_p) = L_p$ and $C_M(t(H^A(X - Y))) \neq L_p$;

(d) $Z_p = X_p - (t(H^A(X - Y)) \cap X_p)$ is a set of support of $X_p$.

By (c), (d), and Theorem 5.1, there is a $(X_p, Z_p)$-supported refutation of $X_p$ in $t(Rs)$ (and, hence, of $t(H^A(X))$). Let us look at the refutation $(T_H)$ of $t(H^A(X))$ constructed in the proof of Theorem 7.7. Clearly, $(T_H)$ is $(H^A(X), Z)$-supported. Moreover, the refutation $(T)$ of $X$, obtained from $(T_H)$ in the way described in the proof of Theorem 7.7, is $(X, Y)$-supported. Indeed, if $\beta_i$ were the resolvent of $\beta_{i_0}, \ldots, \beta_{i_n} \in X - Y$, then the corresponding formulas $\alpha_{i_0}, \ldots, \alpha_{i_n}$ of $(T_H)$, as non-creative instances of $\beta_{i_0}, \ldots, \beta_{i_n}$, would be in $H^A(X - Y)$, which is impossible. ∎

The generalization of the polarity strategy requires a little more work. We discuss the *vo*-polarity only. Let $t$ be an arbitrary but fixed lifting function for $\mathcal{P}$. By Theorem 7.9, $t(Rs)$ is a resolution counterpart of $\mathcal{P}_p$. Let $\prec$ be a binary relation on the set of verifiers of $t(Rs)$ consistent with $\leq_{t(Rs)}$. Because of the 1-1 correspondence between the verifiers of $Rs$ and the verifiers of $t(Rs)$, we can view $\prec$ as a binary relation on the set of verifiers of $Rs$. As in Chapter 5, we assume the existence of two verifiers $v_F, v_T$ of $Rs$ such that $v_T \in C(\emptyset)$ and $C(v_F) = L$ (cf. Exercise 7.4).

To generalize Theorem 5.23 to $FFO$ logics, we only need to define the notion of polarity in a way that does not interfere with Lifting Lemma. The following definition suffices.

*An occurrence of $\beta$ in a formula $\alpha$ of $\mathcal{P}$ is positive (is negative) iff the corresponding occurrence of $t(\beta)$ in $t(\alpha)$ is positive (is negative).*

We can now define the polarity strategy (pol1) for $FFO$ logics in the way analogous to the propositional case:

(pol1) *The resolvent $\kappa(\beta_0)(\kappa(\phi_0)/v_0) \vee \ldots \vee \kappa(\beta_n)(\kappa(\phi_n)/v_n)$ of $\beta_0(\phi_0), \ldots, \beta_n(\phi_n)$ cannot be formed and used during the deductive process if there are $i, j \leq n$ such that $v_i \leq_{Rs} v_j$ and either*

- *$\kappa(\phi_0)$ is positive in $\kappa(\beta_j)$ but not in $\kappa(\beta_i)$*

*or*

- *$\kappa(\phi_0)$ is negative in $\kappa(\beta_i)$ but not in $\kappa(\beta_j)$.*

This definition ensures that any (pol1)-restricted refutation in $t(Rs)$ can be lifted to a (pol1)-restricted refutation in $Rs$. Hence, the following fact holds.

THEOREM 7.12: *Rs admits* (pol1).

The generalization of (pol2) to *FFO* logics is left to the reader. The reader may also check that the combination of either of the polarity strategies with the set of support strategy preserves refutational completeness.

We illustrate the application of the theorem proving strategies discussed in this chapter with the following example.

EXAMPLE 7.13: Let $\mathcal{L}$ be the first-order language obtained from the language described in Example 7.3 by the addition of one logical constant **1**. Let $\mathcal{P}_s^* = \langle \mathcal{L}, C_s^* \rangle$ be the *FFO* logic defined by the Słupecki matrix $M_s$ described in Example 4.9. Clearly, the propositional logic associated with $\mathcal{P}_s^*$ is the Słupecki logic $\mathcal{P}_s$. The resolution counterpart $Rs_3^s$ of $\mathcal{P}_s$, described in Appendix F, has three verifiers:

$$v_2 = \mathbf{1} \to \mathbf{1}, \qquad v_1 = \mathbf{1}, \qquad v_0 = \neg v_2.$$

Since the verifiers of $Rs_3^s$ do not contain propositional variables, by Theorem 7.7, $Rs_3^s$ is also a resolution counterpart of $\mathcal{P}_s^*$. In addition to the 'ground' transformation rules of $Rs_3^s$, it is useful to include more general transformation rules that correspond to polynomial equations true in the algebra of verifiers of $Rs_3^s$; some sample rules are:

$$\begin{aligned}
&\alpha \to v_2 \Rightarrow v_2,\\
&v_0 \to \alpha \Rightarrow v_2,\\
&v_2 \to \alpha \Rightarrow \alpha,\\
&v_0 \vee \alpha \Rightarrow \alpha,\\
&\alpha \vee \alpha \Rightarrow \alpha.
\end{aligned}$$

Since $v_0 =[Rs_3^s]\ v_1$, we let $\prec$ be the relation on the set of verifiers of $Rs_3^s$ defined by: $v_0 \prec v_1 \prec v_2$ and $v_0 \prec v_2$. We may easily verify that for the *vo*-polarity:

$\vee$ and $\wedge$ are positive over both their arguments;

$\to$ is negative over its first and positive over its second argument;

$\neg$ is negative over its only argument.

Now, we use $Rs_3^s$ to refute the set $X$ consisting of the following formulas:

(1) $P(x)^{\ominus} \rightarrow Q(f(x))^{\oplus}$,

(2) $\neg Q(x)^{\ominus} \vee R(x)^{\oplus}$,

(3) $(P(x)^{\oplus} \rightarrow R(x)^{\ominus}) \rightarrow \neg(P(x)^{\oplus} \rightarrow R(x)^{\ominus})$.

In (1), (2), and (3) we have used '$\oplus$' and '$\ominus$' to indicate the polarity of an occurrence of an atomic subformula. So, according to this convention, the only occurrence of $P(x)$ in (1) is negative while the occurrence of $Q(f(x))$ is positive. Initially, there are 24 possible applications of the resolution rule to these formulas. If we choose $\{(3)\}$ as the set of support, then the set of support strategy blocks 10 of these applications, many of which lead to tautologies. The (pol1) strategy blocks 8 of the remaining 14 applications while (pol2) blocks 10 of the 14 applications (the two extra applications blocked by (pol2) involve self-resolving of (3)). Thus the combination of (pol1) with the set of support strategy leaves us with 6 possible initial applications of the resolution rule, and the combination with (pol2) leaves us with only 4 such applications.

In the refutation presented below, we leave out obvious applications of transformation rules.

(4) $(v_1 \rightarrow Q(f(x))) \vee Q(f(x))$, (resolvent of (3),(1),(1))

(5) $\neg Q(x) \vee v_1$, (resolvent of (2),(2),(3))

(6) $v_1$. (resolvent of (4),(5),(5))

This refutation is $(X, \{(3)\})$-supported and consistent with (pol2).

While (pol2) restricts the resolution rule more severely than (pol1) does, we need to deal with the two extra rules $R^+$ and $R^-$. If applied carelessly, these rules can generate tautologies, which are useless from the point of view of the deductive process. For example, $R^+$ can be applied to the formula $q \rightarrow p$ to obtain $q \rightarrow v_2$, which is a tautology. However, in some cases the judicious use of these rules can save space and time by leading to shorter proofs. For example, if we try to refute

(7) $\neg(P(x) \rightarrow (Q(y)^{\oplus} \rightarrow P(x)))$,

then (pol2) strategy will not allow us to self-resolve (7) upon $Q(y)$. Rather, we can use $R^+$ to infer

(8) $\neg(P(x) \to (v_2 \to P(x)))$

which reduces to

(9) $\neg(P(x)^{\oplus} \to P(x)^{\ominus})$.

(9) can be self-resolved to complete a refutation. By using $R^+$ we avoid one resolution step and a number of reductions which would be forced by this additional application of the resolution rule. ∎

## 7.5 RESOLUTION CIRCUITS FOR FFO LOGICS

We adopt the definition of a resolution circuit introduced in Chapter 6 to $FFO$ logics without any significant changes. In this section we show that Theorems 6.7 and 6.8 hold for $FFO$ logics.

Let $\mathcal{P} = \langle \mathcal{L}, C \rangle$ be an arbitrary but fixed $M$-valued $FFO$ logic. Moreover, let $\mathcal{P}_p$ be the propositional logic associated with $\mathcal{P}$ and let $t$ be a lifting function for $\mathcal{P}$. Let us recall from the preceding sections that if $Rs$ is a resolution proof system on $\mathcal{L}$, then $t(Rs)$ denotes the resolution proof system on $\mathcal{L}_p$ obtained from $Rs$ by replacing every verifier $v$ of $Rs$ by $t(v)$. Similarly, for every resolution proof system $Rs_p$ on $\mathcal{L}_p, t^{-1}(Rs_p)$ denotes the resolution proof system on $\mathcal{L}$ obtained from $Rs_p$ by replacing every verifier $v^p$ of $Rs_p$ by $t^{-1}(v^p)$ (we assume, of course, that $t^{-1}(v^p)$ is a ground atomic formula).

THEOREM 7.14: *Suppose that $\langle \mathcal{R}_p, \alpha, f_p \rangle$ is a resolution circuit for $\mathcal{P}_p$. Then $\langle \{t^{-1}(Rs_p) : Rs_p \in \mathcal{R}_p\}, \alpha, f \rangle$ is a resolution circuit for $\mathcal{P}$, where $f(r) = t^{-1}(f_p(r))$, all $r \in Var(\alpha)$.*

*Proof:* Let $A$ be the set of all constant symbols of $\mathcal{L}$ that do not occur in verifiers of $t^{-1}(Rs_p), Rs_p \in \mathcal{R}_p$. We assume that $A \neq \emptyset$. For every $Rs_p \in \mathcal{R}_p$ let $C_{Rs_p}$ denote the consequence operation defined by the matrix $M_{Rs_p}$ induced by $Rs_p$. Moreover, let $\langle \mathcal{L}, C^*_{Rs_p} \rangle$ be the $FFO$ logic defined by the matrix $M_{Rs_p}$. Let $X \subseteq L$ be finite and clean in every $t^{-1}(Rs_p), Rs_p \in \mathcal{R}_p$. By Herbrand's Theorem and the hypothesis,

(a) $C(X) = L$ *iff* $C_M(t(H^A(X))) = L_p$ *iff* $\alpha[r/t(H^A(X)) \overset{f(r)}{\Rightarrow} \Box]$ *is true.*

So, to complete the proof of this theorem we need to show that for every $Rs_p \in \mathcal{R}_p$,

(b) $t(H^A(X)) \overset{Rs_p}{\Longrightarrow} \Box$ *iff* $X \overset{t^{-1}(Rs_p)}{\Longrightarrow} \Box$.

Let $Rs_p \in \mathcal{R}_p$. Then $t(H^A(X)) \overset{Rs_p}{\Longrightarrow} \Box$ *iff* $C_{Rs_p}(t(H^A(X))) = L_p$, which, by Herbrand's Theorem, gives us

(c) $t(H^A(X)) \overset{Rs_p}{\Longrightarrow} \Box$ *iff* $C^*_{Rs_p}(X) = L$.

By Theorem 7.7,

(d) $C^*_{Rs_p}(X) = L$ *iff* $X \overset{t^{-1}(Rs_p)}{\Longrightarrow} \Box$.

So, (b) follows from (c) and (d). ∎

THEOREM 7.15: *Suppose that* $\langle \mathcal{R}, \alpha, f \rangle$ *is a resolution circuit for* $\mathcal{P}$. *Then* $\langle \{t(Rs) : Rs \in \mathcal{R}\}, \alpha, f_p \rangle$ *is a resolution circuit for* $\mathcal{P}_p$*, where* $f_p(r) = t(f(r))$*, all* $r \in Var(\alpha)$.

*Proof:* Let $X \subseteq L_p$ be finite and clean in every $t(Rs)$, $Rs \in \mathcal{R}$. By Proposition 1.4, we can assume that $t^{-1}(X)$ consists of ground formulas only. For every $Rs \in \mathcal{R}$, let $N_{Rs}$ be a finite matrix for $\mathcal{L}_p$ such that $Rs$ is a resolution counterpart of the $N_{Rs}$-valued $FFO$ logic $\langle \mathcal{L}, C^*_{N_{Rs}} \rangle$. So,

(a) $C^*_{N_{Rs}}(t^{-1}(X)) = L$ *iff* $t^{-1}(X) \overset{Rs}{\Longrightarrow} \Box$.

By Theorem 7.9, $t(Rs)$ is a resolution counterpart of $\langle \mathcal{L}_p, C_{N_{Rs}} \rangle$. So,

(b) $C_{N_{Rs}}(X) = L_p$ *iff* $X \overset{t(Rs)}{\Longrightarrow} \Box$.

Moreover, by Herbrand's Theorem,

(c) $C^*_{N_{Rs}}(t^{-1}(X)) = L$ *iff* $C_{N_{Rs}}(X) = L_p$.

From (a), (b), and (c) we obtain

(d) $t^{-1}(X) \overset{R_s}{\Rightarrow} \square$ *iff* $X \overset{t(R_s)}{\Longrightarrow} \square$.

Now, by Herbrand's Theorem and the hypothesis,

$$C_M(X) = L_p \text{ iff } C(t^{-1}(X)) = L \text{ iff } \alpha[r/t^{-1}(X) \overset{f(r)}{\Longrightarrow} \square] \text{ is true},$$

which, by (d), gives us $C_M(X) = L_p$ *iff* $\alpha[r/X \overset{f_p(r)}{\Longrightarrow} \square]$ *is true.* ∎

An immediate consequence of Theorems 6.7, 6.8, 7.14, and 7.15 is the fact that an optimal resolution circuit for an $FFO$ logic $\mathcal{P}$ can be effectively constructed from an optimal resolution circuit for the propositional logic associated with $\mathcal{P}$, in the way described in Theorem 7.14.

**EXERCISES**

7.1. (Hawranek and Tokarz [20]) Let $\mathcal{P} = \langle \mathcal{L}, C \rangle$ be an $M$-valued $FFO$ logic and suppose that $\mathcal{L}$ is without function symbols (i.e., $F_L = \emptyset$). Show that $\mathcal{P}$ is compact, i.e., that $\mathcal{P}$ satisfies the definition of compactness formulated for propositional logics in Section 1.3.

7.2. Let $\mathcal{P} = \langle \mathcal{L}, C \rangle$ be an $FFO$ logic and let $f_0, \ldots, f_k$ be the logical connectives of $\mathcal{L}$. Let $\Theta_C^R$ be the binary relation on $L$ define as follows: for every $\alpha, \beta \in L$,

$$\alpha\, \Theta_C^R\, \beta \text{ iff } C(X, \gamma(A/\alpha)) = L \Leftrightarrow C(X, \gamma(A/\beta)) = L, \text{ for every } X \cup \{\gamma(A)\} \subseteq L,$$

where $A$ is an atomic subformula of $\gamma$. Show that $\Theta_C^R$ is a congruence of the algebra $\langle \mathcal{L}, f_0, \ldots, f_k \rangle$.

7.3. Let $\langle \mathcal{L}, C \rangle$ be an $M$-valued $FFO$ logic, let $Y, X \subseteq L$, and let $\kappa$ be a substitution. Show that $C(Y \cup X) \neq L$ implies $C(Y \cup \kappa(X)) \neq L$.

7.4. Let $\mathcal{P} = \langle \mathcal{L}, C \rangle$ be an $M$-valued $FFO$ logic and let $t$ be a lifting function for $\mathcal{P}$. Show that for every ground formula $\alpha \in L, \alpha \in C(\emptyset)$ *iff* $t(\alpha) \in C_M(\emptyset)$.

7.5. Let $\mathcal{P}$ be an $FFO$ logic with infinitely many ground atomic formulas. Show that $\mathcal{P}$ has a resolution counterpart $Rs$ such that at most one constant symbol occurs in the verifiers of $Rs$.

7.6. Let $Rs$ be a resolution counterpart of an $M$-valued $FFO$ logic $\mathcal{P}$ and let $t$ be a lifting function for $\mathcal{P}$. Show that $Rs$ is a resolution counterpart of the $FFO$ logic defined by the matrix induced by $t(Rs)$.

7.7. Prove an analog of Theorem 6.10 for $FFO$ logics.

# 8

# NONMONOTONIC RESOLUTION INFERENCE SYSTEMS

Nonmonotonic inference systems, i.e., inference systems that fail to satisfy the monotonicity principle (c2), arose as a result of the search for logical tools which can handle a variety of forms of reasoning involving inferences based on information which can be incomplete, uncertain, approximate, or subject to revision. In this chapter we show one possible way of extending the resolution proof system methodology developed in the previous chapters to cumulative nonmonotonic inference systems.

## 8.1 CUMULATIVE INFERENCE SYSTEMS

Let $\mathcal{L}$ be a propositional language. We say that a function $C : 2^L \to 2^L$ is an *inference operation* on $\mathcal{L}$ if it satisfies the condition (c1) of inclusion and (c3) of idempotence formulated in Section 1.1. If $C$ is an inference operation on $\mathcal{L}$, then $\langle \mathcal{L}, C \rangle$ is called an *inference system*. Nonmonotonic inference systems, i.e., inference systems that violate the principle (c2) of monotonicity, contrast sharply with the inference systems we have discussed so far. By replacing monotonicity with some other principles of reasoning we challenge many properties of inference, traditionally considered natural or useful. In a nonmonotonic context, the addition of new premises can lead to the retraction of previous conclusions. As a result, theories (i.e., inferentially closed sets of formulas) lose their prominent role which they enjoyed in the analysis of logical calculi. As shown in [70], nonmonotonic theories alone are not sufficiently expressive for the analysis of nonmonotonic inferences.

One way of extending our theory of resolution proof systems to nonmonotonic inference systems is to look for possible links between nonmonotonic calculi and resolution logics. In Sections 1.2 and 3.5, we argued that the search for an efficient automated reasoning algorithm for a logic $\mathcal{P}$ should include the study of other logics in the resolution class $\mathcal{K}_{\mathcal{P}}$ of $\mathcal{P}$. If $\mathcal{P}_N$ is a nonmonotonic inference system on $\mathcal{L}$, then we can also inspect the class $\mathcal{K}_{\mathcal{P}_N}$ of all resolution logics on $\mathcal{L}$ which have the same inconsistent sets of formulas as $\mathcal{P}_N$, looking for a suitable refutational variant. Clearly, every resolution counterpart of a logic in $\mathcal{K}_{\mathcal{P}_N}$ can be used to determine the consistency status of finite sets of formulas of $\mathcal{P}_N$.

It should be stressed, however, that this approach will not work for every inference system. It suffices to point out that the refutational principle is not universally available to all inference systems, as it is not available to all logics. Furthermore, there are nonmonotonic inference systems $\mathcal{P}_N$ for which no logic with the same inconsistent sets of formulas as $\mathcal{P}_N$ exists. This happens exactly when $\mathcal{P}_N = \langle \mathcal{L}, C \rangle$ violates the following property: for every $X, Y \subseteq L$,

(c4) $$\text{if } X \subseteq Y \text{ and } C(X) = L, \text{ then } C(Y) = L.$$

In other words, (c4) is the necessary and sufficient condition for an inference system to have the same inconsistent sets of formulas as some logical (i.e., monotonic) calculus (cf. Exercise 8.1).

Cumulative inference systems, which have risen to prominence as a result of the search for desired and natural formal properties of nonmonotonic reasoning, are among the inference operations that satisfy (c4). An inference operation $C$ is said to be *cumulative* if it satisfies the following *principle of cumulativity*: for every $X, Y \subseteq L$,

(c5) $$X \subseteq Y \subseteq C(X) \text{ implies } C(Y) = C(X).$$

Clearly, (c1) and (c5) imply (c3). Since (c5) implies (c4), for every cumulative inference system $\mathcal{P}_N$, there exists a logic $\mathcal{P}$ with the same inconsistent sets of formulas as $\mathcal{P}_N$. (However, we cannot guarantee that $\mathcal{P}$ is a resolution logic.) The reader may refer to [27, 28] for the comprehensive presentation of the theory of cumulative inference systems.

In Sections 8.2 and 8.3, we introduce several notions which are required to state and prove the main results of this chapter. These results, which are collected in Section 8.4, extend the range of application of the theory of resolution proof systems to (resolution) cumulative inference systems.

## 8.2 PREFERENTIAL MATRICES

Matrix semantics has played a central role in the development of the theory of resolution proof systems presented in the previous chapters. This role is exemplified, for example, by the notion of the matrix induced by a resolution proof system. These matrices provide an important methodological link between resolution proof systems and resolution logics. An extension of our theory of resolution proof systems beyond the class of resolution logics, however, calls for a revision of the notion of a logical matrix. Needless to say, the inference operation defined by a logical matrix is monotonic and structural while nonmonotonic formalisms depart not only from monotonicity but frequently from structurality as well.

We begin this revision process by briefly reviewing the semantic studies of nonmonotonic inference systems undertaken in [6, 23, 27, 28]. All these works share a common preferential model-theoretic view on semantics. In [27, 28] this unified semantic framework assumes the form of the theory of preferential model structures. If $\mathcal{L}$ is the language of an inference system, then a ***preferential model structure*** for $\mathcal{L}$ is a triple

$$M = \langle \mathcal{U}, \models, \prec \rangle,$$

where:

- $\mathcal{U}$ is a nonempty set; the elements of $\mathcal{U}$ are called *models*;
- $\models$ is a binary relation between models in $\mathcal{U}$ and formulas of $\mathcal{L}$ called the *satisfaction relation* of $M$;
- $\prec$ is a binary relation on $\mathcal{U}$ called the *preference relation* of $M$.

No properties of $\mathcal{U}$, of the satisfaction relation $\models$, nor of the preference relation $\prec$ are assumed. Makinson proved that, under a suitable definition of the operation of preferential entailment (cf. Exercise 8.2), cumulative inference operations are exactly those defined by the class of (stoppered) preferential model structures (cf. [28]).

The study of general properties of cumulative inference systems can be based on a less general and more structured notion of a model structure called a *preferential matrix* (cf. [68]). As in the case of logical matrices, the key feature of the preferential matrix proposal is the truth-functional interpretation of logical connectives. We begin our voyage towards the notion of a preferential matrix by analyzing the classical notion of a logical matrix reviewed in Section 1.4.

Let $\mathcal{L}$ be an arbitrary propositional language fixed for the remainder of this chapter. Consequence operations defined by logical matrices for $\mathcal{L}$ are always monotonic and structural. As we have mentioned above, nonmonotonic formalisms are frequently non-structural. One way of extending matrix semantics to cover all consequence operations, and not just structural once, is described in [49, 60]. The key idea is to base the semantic entailment on a set of 'admissible valuations', i.e., to consider *generalized matrices* of the form $\langle \mathcal{A}, \mathcal{D}, \mathcal{H} \rangle$, where $\langle \mathcal{A}, \mathcal{D} \rangle$ is a logical matrix for $\mathcal{L}$ and $\mathcal{H}$ is a subset of the set of all valuations of $\mathcal{L}$ into $\langle \mathcal{A}, \mathcal{D} \rangle$.

With every generalized matrix $M$ we associate the consequence operation $C_M$ defined as follows: for every $X \cup \{\alpha\} \subseteq L$,

(cs) $\alpha \in C_M(X)$ *iff* for every $h \in \mathcal{H}$ and every $d \in \mathcal{D}$, $h(X) \subseteq d$ implies $h(\alpha) \in d$.

In this semantic framework, every consequence operation can be defined by a generalized matrix. In [23], Kraus, Lehmann and Magidor used a similar idea (of restricting the set of possible interpretations) to go beyond structural nonmonotonic inference systems.

Our main problem with semantic modeling of cumulative inference systems, however, is not structurality but monotonicity. To get over this problem, we employ the idea of a *preference relation* so successfully used in preferential model-theoretic semantics which we have briefly reviewed at the beginning of this section. We call a system

$$M = \langle \mathcal{A}, \mathcal{D}, \mathcal{H}, \sqsubset \rangle$$

a *preferential matrix* for $\mathcal{L}$ if $\langle \mathcal{A}, \mathcal{D}, \mathcal{H} \rangle$ is a generalized matrix and $\sqsubset$ is a binary relation on $\mathcal{D}$. We call $\sqsubset$ the *preference relation* of $M$ and for every pair $d_0, d_1 \in \mathcal{D}$, we read '$d_0 \sqsubset d_1$' as '$d_0$ is preferred over $d_1$'.

Given a preferential matrix $M = \langle \mathcal{A}, \mathcal{D}, \mathcal{H}, \sqsubset \rangle$, the notion of preferential satisfaction is defined in the following way. If $X \subseteq L$, $d \in \mathcal{D}$, and if $h \in \mathcal{H}$, then $X$ is *preferentially satisfied* in $M$ by $h$ and $d$, written as $Sat_M(h, X, d)$, if and only if $h(X) \subseteq d$ and for every $d' \sqsubset d$, $h(X) \not\subseteq d'$. Intuitively, $Sat_M(h, X, d)$ means that $d$ is a most preferred set of designated truth-values containing $h(X)$.

Next, we use the notion of preferential satisfaction to associate an inference operation $C_M$ with $M$ by rewriting the definition (cs) in the following way: for every $X \cup \{\alpha\} \subseteq L$,

$\alpha \in C_M(X)$ *iff* for every $h \in \mathcal{H}$ and every $d \in \mathcal{D}$, $Sat_M(h, X, d)$ implies $h(\alpha) \in d$.

We call a preferential matrix $M = \langle \mathcal{A}, \mathcal{D}, \mathcal{H}, \sqsubset \rangle$ *stoppered* if and only if for every set $A$ of truth-values of $M$, the set $\mathcal{Y}_A = \{d \in \mathcal{D} : A \subseteq d\}$ is empty or has the smallest element, i.e., there is $d_A \in \mathcal{Y}_A$ such that for every $d \in \mathcal{Y}_A, d \neq d_A$ implies $d_A \sqsubset d$; moreover, for no $d \in \mathcal{Y}_A, d \sqsubset d_A$. We illustrate the definition of a stoppered preferential matrix with the following example.

EXAMPLE 8.1: Let $\langle \mathcal{L}, C \rangle$ be a cumulative inference system, let $id$ be the identity function on $L$, and let $\sqsubset$ be the binary relation on the set $Th(C) = \{C(X) : X \subseteq L\}$ of theories of $C$ defined in the following way: for every $X, Y \in Th(C)$,

$$X \sqsubset Y \text{ iff } X \neq Y \text{ and for some } Y_0 \subseteq Y,\ X = C(Y_0).$$

We call the preferential matrix

$$\langle \mathcal{L}, Th(C), \{id\}, \sqsubset \rangle$$

the *Lindenbaum preferential matrix* for $C$. It is not difficult to verify that $M$ is stoppered and that $C = C_M$.

To prove that $M$ is stoppered, let $A \subseteq L$ and let $\mathcal{Y}_A = \{X \in Th(C) : A \subseteq X\}$. Clearly, $\mathcal{Y}_A \neq \emptyset$. Let $A^* = \bigcap \mathcal{Y}_A$. So, $C(A^*) \in \mathcal{Y}_A$. We claim that $C(A^*)$ is the smallest element of $\mathcal{Y}_A$. To prove the claim, let $C(X) \neq C(A^*)$ be any element of $\mathcal{Y}_A$. Since $A^* \subseteq C(X)$, we have $C(A^*) \sqsubset C(X)$. Finally, if for some $C(X) \in \mathcal{Y}_A, C(X) \sqsubset C(A^*)$, then there exists $X' \subseteq C(A^*)$ such that $C(X) = C(X')$. Hence, $A^* \subseteq C(X')$ and, by cumulativity of $C$, $C(X) = C(A^*)$. This, however, contradicts the assumption that $C(X) \sqsubset C(A^*)$. Hence, $C(A^*)$ is the smallest element of $\mathcal{Y}_A$, as claimed.

To show that $C = C_M$, let us first verify that

(a) for every $X \subseteq L, Sat_M(id, X, C(X))$.

To show (a), suppose that there is $Y \subseteq L$ such that $C(Y) \sqsubset C(X)$ and $id(X) \subseteq C(Y)$. Then, by the definition of $\sqsubset$, there exists $Y' \subseteq C(X)$ such that $C(Y) = C(Y')$.

Hence, $X \subseteq C(Y')$ and $Y' \subseteq C(X)$ which, by inclusion and cumulativity, gives us $C(X) = C(Y)$. This, of course, contradicts the assumption that $C(Y) \sqsubset C(X)$.

In a similar way we can prove that

(b) for every $X, Y \subseteq L$, if $Sat_M(id, X, C(Y))$, then $C(X) = C(Y)$.

Indeed, if $Sat_M(id, X, C(Y))$ and $C(X) \neq C(Y)$, then $X \subseteq C(Y)$ and we must have $C(X) \sqsubset C(Y)$. This, however, contradicts the choice of $C(Y)$.

Finally, from (a) and (b) we immediately conclude that $C = C_M$. ∎

THEOREM 8.2: *An inference operation $C$ is cumulative iff there is a preferential stoppered matrix $M$ such that $C = C_M$.*

*Proof:* In Example 8.1 we demonstrated that every cumulative inference operation $C$ on $\mathcal{L}$ is defined by the preferential (and stoppered) Lindenbaum matrix for $C$.

Now, let $M = \langle \mathcal{A}, \mathcal{D}, \mathcal{H}, \sqsubset \rangle$ be a preferential stoppered matrix for $\mathcal{L}$. We want to show that $C_M$ is a cumulative inference operation on $\mathcal{L}$. Clearly, $C_M$ satisfies inclusion. To show that $C_M$ is cumulative, let $X, Y \subseteq L$ be such that $X \subseteq Y \subseteq C_M(X)$. Let us assume that $\alpha \in C_M(X)$ and that for some $d \in \mathcal{D}$ and some $h \in \mathcal{H}, Sat_M(h, Y, d)$. So, $h(X) \subseteq d$ and, since $M$ is stoppered, for some $d^* \in \mathcal{D}, Sat_M(h, X, d^*)$. Clearly, $d = d^*$. Indeed, if $d^* \neq d$, then, since $h(C_M(X)) \subseteq d^*$, we would have $h(Y) \subseteq d^* \sqsubset d$, which would contradict the choice of $d$. Hence, $h(\alpha) \in h(C_M(X)) \subseteq d$ and $\alpha \in C_M(Y)$, as required.

Conversely, if $\alpha \in C_M(Y)$ and for some $h \in \mathcal{H}$ and $d \in \mathcal{D}, Sat_M(h, X, d)$, then $h(Y) \subseteq h(C_M(X)) \subseteq d$. If for some $d^* \sqsubset d, h(Y) \subseteq d^*$, then we would have $h(X) \subseteq d^*$. This, of course, is impossible by the choice of $d$. Hence, $Sat_M(h, Y, d)$ and, since $\alpha \in C_M(Y), h(\alpha) \in d$. We therefore conclude that $\alpha \in C_M(X)$, as required. ∎

We say that an inference operation $C$ on $\mathcal{L}$ is *structural* if for every $X \subseteq L$ and every substitution $e$, $e(C(X)) \subseteq C(e(X))$.

THEOREM 8.3: *An inference operation $C$ is cumulative and structural iff there is a preferential stoppered matrix $M = \langle \mathcal{A}, \mathcal{D}, \mathcal{H}, \sqsubset \rangle$ such that $C = C_M$ and $\mathcal{H}$ consists of all valuations.*

*Proof:* Let $C$ be a cumulative inference operation on $\mathcal{L}$. Consider the preferential matrix $M = \langle \mathcal{L}, \{C(X) : X \subseteq L\}, \mathcal{H}, \sqsubset \rangle$ for $\mathcal{L}$, where $\sqsubset$ is defined as for the preferential Lindenbaum matrix for $C$ and $\mathcal{H}$ is the set of all logical substitutions. We ask the reader to verify that $M$ is stoppered and defines $C$.

Now, suppose that $C$ is defined by some stoppered preferential matrix $M = \langle \mathcal{A}, \mathcal{D}, \mathcal{H}, \sqsubset \rangle$ such that $\mathcal{H}$ consists of all valuations. By Theorem 8.2, $C_M$ is a cumulative inference operation. To show the structurality of $C_M$, let $X \cup \{\alpha\} \subseteq L$ be such that $\alpha \in C_M(X)$ and let $e$ be an arbitrary substitution. If $h \in \mathcal{H}$ and $d \in \mathcal{D}$ are such that $Sat_M(h, e(X), d)$, then $h(e(X)) = e \circ h(X) \subseteq d$. Clearly, the composition $e \circ h$ is in $\mathcal{H}$. We claim that

(a) $Sat_M(e \circ h, X, d)$.

Indeed, if for some $d^* \in \mathcal{D}$, $d^* \sqsubset d$ and $e \circ h(X) \subseteq d^*$, then $h(e(X)) \subseteq d^*$. This, of course, contradicts the assumption that $Sat_M(h, e(X), d)$. From (a) and the fact that $\alpha \in C_M(X)$ we obtain $e \circ h(\alpha) = h(e(\alpha)) \in d$ and conclude that $e(\alpha) \in C_M(e(X))$.∎

There are obvious connections between preferential matrices and logical matrices (clearly, if $\langle \mathcal{A}, \mathcal{D}, \mathcal{H}, \sqsubset \rangle$ is a preferential matrix, then $\langle \mathcal{A}, \mathcal{D} \rangle$ is a logical matrix). Hence, one may expect that some of the algebraic tools and techniques developed for logical matrices can be adopted for the study of cumulative inference systems. In this and the following sections we shall try to do just that in the context of finding resolution counterparts of cumulative inference systems.

## 8.3 MONOTONE BASES OF INFERENCE OPERATIONS

Nonmonotonic inference systems are usually formulated with respect to, or 'on the basis of', some monotonic logical systems; they depart from their deductive (i.e., monotonic) counterparts by giving up monotonicity for some other principles of inference. Nonmonotonic inference systems of Poole (cf. [28, 50]) are based on classical logic and so are the so-called supraclassical systems discussed by Makinson in [28]. Frequently, nonmonotonic inference systems are based on non-classical logics: on Kleene's three-valued logic (cf. [11]), on modal logics (cf. [31, 34, 36]), on constructive logic (cf. [47]), etc. This suggests the following definition. The *monotone base* of a cumulative inference operation $C$ is the greatest structural consequence operation $C_B$ such that $C_B \leq C$ (the greatest with respect to $\leq$).

THEOREM 8.4: *Every cumulative inference operation $C$ has the unique monotone base $C_B$. $C_B$ is defined by the matrix $N = \langle \mathcal{L}, \{C(X) : X \subseteq L\}\rangle$.*

*Proof:* Let $C$ and $N$ be as stated and let $M$ be the preferential Lindenbaum matrix for $C$. We want to show that:

(a) $C_N$ is structural and $C_N \leq C$;

(b) for every structural consequence $C^*$ on $\mathcal{L}$, $C^* \leq C$ implies $C^* \leq C_N$.

From (a) and (b) our theorem follows immediately.

Since $N$ is a logical matrix, $C_N$ is structural. To show that $C_N \leq C$, let $X \cup \{\alpha\} \subseteq L$ be such that $\alpha \in C_N(X)$ and let $Y \subseteq L$ be such that $Sat_M(id, X, C(Y))$. Since $id$ is a valuation of $\mathcal{L}$ into $N$ and since $C(Y)$ is a set of designated truth-values of $N$, $id(\alpha) \in C(Y)$. Hence, $\alpha \in C_M(X)$ and (a) has been justified.

To show (b), let $C^* \leq C$ be an arbitrary structural consequence operation on $\mathcal{L}$. By (c1) and (c3), for every $X \subseteq L$, $C^*(C(X)) = C(X)$. So,

(c) $\{C(X) : X \subseteq L\} \subseteq \{C^*(X) : X \subseteq L\}$.

Now, (b) follows from (c) and from the fact that $C_N$ is defined by $N$ while $C^*$ is defined by $\langle \mathcal{L}, \{C^*(X) : X \subseteq L\}\rangle$ (an easy proof using the structurality of $C^*$ is left to the reader). ∎

In [28], a cumulative inference operation $C$ is called *supraclassical* if $C_2 \leq C$, where $C_2$ denotes the consequence operation of classical logic. If $C$ is nontrivial (i.e., if $C(X) \neq L$, for some $X \subseteq L$), then supraclassicality is equivalent to the statement that $C_2$ is the monotone base of $C$, i.e., that $C_B = C_2$. Indeed, by Theorem 8.4, $C_2 \leq C_B \leq C$. Since $C_2$ is a maximal structural consequence operation, i.e., for every structural consequence operation $C^* \geq C_2$, $C^* = C_2$ or $C^*(\emptyset) = L$, and since $C$ is nontrivial, we must have $C_B = C_2$.

## 8.4 CONSISTENCY PRESERVATION

Theorem 8.4 states that every cumulative inference operation $C$ on $\mathcal{L}$ has the unique monotone base $C_B$. If $\langle \mathcal{L}, C_B\rangle$ is a resolution logic and if this logic has the same inconsistent sets of formulas as $\langle \mathcal{L}, C \rangle$, then every resolution counterpart $Rs$ of $\langle \mathcal{L}, C_B \rangle$ satisfies (r1)–(r3) with respect to $C$. In other words, $Rs$ becomes a resolution counterpart of $\langle \mathcal{L}, C \rangle$. This is exactly the situation we investigate in this section.

We begin with the following definition. We say that an inference operation $C$ satisfies the *consistency preservation property* if for every $X \subseteq L$,

$$C(X) = L \text{ } \textit{iff} \text{ } C_B(X) = L.$$

In other words, this property, when satisfied, sets the limit on how much a given inference operation $C$ and its monotone base $C_B$ may differ. In the context of supraclassical inference systems, the consistency preservation property was investigated in [28].

> If cumulativity can be thought of as a principle for organizing our conclusions coherently, and supraclassicality and distribution as principles giving them strength, then consistency preservation can be seen as a condition limiting the power of our inferences: $C$ never takes us into contradictions that were not already classically implicit in our premises. [28]

For a list of inference systems with and without the consistency preservation property the reader is referred to [28].

The consistency preservation property can be derived from cumulativity, structurality, and the property of inference systems which we call *properness*. An inference operation $C$ is said to be *proper* if and only if $C(e(L)) = L$, for every substitution $e$ (cf. Example 8.3).

THEOREM 8.5: *Every cumulative, structural, and proper inference operation satisfies the consistency preservation property.*

*Proof:* Let $C$ be an inference operation on $\mathcal{L}$ which satisfies the assumptions of our theorem and let $M$ be the preferential Lindenbaum matrix for $C$. By Theorem 8.3, we can assume that the set of valuations of $M$ consists of all endomorphisms of $\mathcal{L}$. Since $C_B \leq C$, we must only show that every $C_B$-consistent set is also $C$-consistent. To this end, let $X \subseteq L$ be such that $C_B(X) \neq L$. By Theorem 8.4, $C_B$ is defined by the logical matrix $\langle \mathcal{L}, \{C(X) : X \subseteq L\} \rangle$. Hence, for some endomorphism $e$ of $\mathcal{L}$ and some $Y \subseteq L$,

(a) $e(X) \subseteq C(Y) \neq L$.

We claim that

(b) $e(L) \not\subseteq C(e(X))$.

Indeed, if (b) were false, then, since $C(e(L)) = L$, we would have $e(L) \subseteq C(e(X)) \subseteq C(e(L))$. By cumulativity and idempotence, this would imply $C(e(X)) = L$. So, by (a), $e(X) \subseteq C(Y) \subseteq C(e(X))$, which, again by (c3) and (c5), would give us $C(Y) = C(e(X)) = L$, contradicting (a). By (b), there is $\alpha \in L$ such that $e(\alpha) \notin C(e(X))$. Since $C$ is structural, $\alpha \notin C(X)$ and the proof of $C$-consistency of $X$ is finished. ∎

The final step of our program for extending the theory of resolution proof systems to cumulative inference systems is to determine which of these calculi can be linked with resolution proof systems. In fact, this can be done with a little effort for cumulative inference systems that satisfy the consistency preservation property.

First, we adopt the definition of a resolution logic to cumulative inference systems. We say that a cumulative inference system $\mathcal{P}_N$ is a *resolution inference system* if and only if there exists a resolution logic with the same inconsistent sets of formulas as $\mathcal{P}_N$.

PROPOSITION 8.6: *Let $C$ be a cumulative inference operation on $\mathcal{L}$ which satisfies the consistency preservation property. Then $\langle \mathcal{L}, C \rangle$ is a resolution inference system iff $\langle \mathcal{L}, C_B \rangle$ is a resolution logic.*

*Proof:* Follows from Theorem 3.19. ∎

COROLLARY 8.7: *Let $C$ be a cumulative and structural inference operation on $\mathcal{L}$ and suppose that $\langle \mathcal{L}, C_B \rangle$ is a resolution logic. Then $\langle \mathcal{L}, C \rangle$ is a resolution inference system.*

*Proof:* Follows from Lemma 3.1(iv), Theorem 8.5, and Proposition 8.6. ∎

In view of Proposition 8.6, it is not difficult to adopt the characterization of the class of resolution logics given in Theorem 3.7 to cumulative inference systems that satisfy the consistency preservation property.

THEOREM 8.8: *Let $C$ be a cumulative inference operation with the consistency preservation property. The following conditions are equivalent:*

(i) *$\langle \mathcal{L}, C \rangle$ is a resolution inference system;*

(ii) *there exists a finite logical matrix $M$ such that $C$ and $C_M$ have the same inconsistent sets;*

(iii) *for some integer $k \geq 0$:*

(a) $L^{(k)}/\Theta^R_{C_B}$ *is finite;*

(b) *for every finite set $X \subseteq L, C(X) = L$ iff for every substitution $e$ which maps $L$ into $L^{(k)}, C(e(X)) = L$.*

*Proof:* Follows from Proposition 8.6 and Theorem 3.7. ∎

We close this chapter with the following example.

EXAMPLE 8.9: In this example we show that every Poole system without constraints (cf. [28, 50]) is a resolution inference system. Our presentation of Poole systems follows [28].

Let $\mathcal{L}$ be the language of classical propositional logic. A *Poole system* determined by a set $\Delta \subseteq L$ of *default propositions* is the inference system $\langle \mathcal{L}, C_\Delta \rangle$, where the inference operation $C_\Delta$ is defined by: for every $Y \subseteq L$,

$$C_\Delta(Y) = \begin{cases} \bigcap Ex(Y), & \text{if } Ex(Y) \neq \emptyset, \\ L, & \text{otherwise,} \end{cases}$$

where

$$Ex(Y) = \{C_2(Y \cup D) : D \textit{ is maximal among the subsets } D' \textit{ of } \Delta \textit{ such that } C_2(Y \cup D') \neq L\}.$$

In [28] it is shown that $C_\Delta$ is cumulative, supraclassical, and that it satisfies the consistency preservation property. Hence, by Proposition 8.6, $\langle \mathcal{L}, C_\Delta \rangle$ is a resolution inference system and the resolution counterpart $Rs_2$ of $\mathcal{P}_2$ can be used to determine the consistency status of finite $C_\Delta$-inconsistent sets of formulas.

The question that remains to be answered is the existence of a refutational principle for Poole systems $\langle \mathcal{L}, C_\Delta \rangle$. In what follows, we give a partial answer to this question for Poole systems with finite sets of defaults. If $\Delta$ is finite, then, of course, $\mathcal{P}_\Delta = \langle \mathcal{L}, C_\Delta \rangle$ is decidable. By Exercise 8.8, $\mathcal{P}_\Delta$ satisfies the refutational principle. The following two facts can be used to compute appropriate deduced sets. Let $X \cup \{\alpha\} \subseteq L$ be finite and such that $X \cup \Delta \cup \{\alpha\}$ is clean in $Rs_2$. Then:

(a) if $X \cup \{\neg\alpha\} \overset{R_{g2}}{\Rightarrow} \Box$, then $\alpha \in C_\Delta(X)$;

(b) if $\alpha \in C_\Delta(X)$, then $X \cup \Delta \cup \{\neg\alpha\} \overset{R_{g2}}{\Rightarrow} \Box$.

To show (a), suppose that $X \cup \{\neg\alpha\} \overset{R_{g2}}{\Rightarrow} \Box$. By (r1), $C_2(X \cup \{\neg\alpha\}) = L$ and, hence, $\alpha \in C_2(X)$. Since $C_2 \leq C_\Delta$, $\alpha \in C_\Delta(X)$.

To show (b) we use the following fact first shown by M. Freund (cf. [28], Observation 3.3.4).

If $\Delta$ is finite, then for every $X \subseteq L, C_\Delta(X) \subseteq C_2(X \cup C_\Delta(\emptyset))$.

Since $C_\Delta(\emptyset) \subseteq C_2(\Delta)$, we have

(c) if $\Delta$ is finite, then for every $X \subseteq L, C_\Delta(X) \subseteq C_2(X \cup \Delta)$.

Now, if $\alpha \in C_\Delta(X)$, then by (c), $\alpha \in C_2(X \cup \Delta)$. So, $C_2(X \cup \Delta \cup \{\neg\alpha\}) = L$ and, by (r1), $X \cup \Delta \cup \{\neg\alpha\} \overset{R_{g2}}{\Rightarrow} \Box$, as required. ∎

## EXERCISES

8.1. Let $C$ be an inference operation on $\mathcal{L}$ and let $\mathcal{K}_C$ be the class of all consequence operations $Cn$ on $\mathcal{L}$ such that for every $X \subseteq L$,

$$C(X) = L \text{ iff } Cn(X) = L.$$

Demonstrate that $\mathcal{K}_C \neq \emptyset$ if and only if $C$ satisfies condition (c4) formulated in Section 8.1.

8.2. Let $M = \langle \mathcal{U}, \models, \prec \rangle$ be a preferential model structure for $\mathcal{L}$. We say that a model $\mathfrak{R} \in \mathcal{U}$ preferentially satisfies a set $X \subseteq L$, written $\mathfrak{R} \models^* X$, if for every $\alpha \in X$, $\mathfrak{R} \models \alpha$ and for every $\mathfrak{I} \in \mathcal{U}$, if $\mathfrak{I} \prec \mathfrak{R}$, then there exists $\alpha \in X$ such that $\mathfrak{I} \not\models \alpha$. With $M$ we associate the operation $C_M : 2^L \longrightarrow 2^L$ of *preferential entailment* defined in the following way. For every $X \cup \{\alpha\} \subseteq L$,

$$\alpha \in C_M(X) \textit{ iff } \text{for every } \mathfrak{R} \in \mathcal{U}, \mathfrak{R} \models^* X \text{ implies } \mathfrak{R} \models \alpha.$$

Let $N = \langle \mathcal{A}, \mathcal{D}, \mathcal{H}, \sqsubset \rangle$ be a preferential matrix for $\mathcal{L}$, and let $M = \langle \mathcal{U}, \models, \prec \rangle$ be a preferential model structure, where:

- $\mathcal{U} = \{\langle \mathcal{A}, h, d \rangle : h \in \mathcal{H}, d \in \mathcal{D}\}$;
- $\langle \mathcal{A}, h_0, d_0 \rangle \prec \langle \mathcal{A}, h_1, d_1 \rangle$ *iff* $h_0 = h_1$ and $d_0 \sqsubset d_1$;
- $\langle \mathcal{A}, h, d \rangle \models \alpha$ *iff* $h(\alpha) \in d$.

Show that $C_M = C_N$.

8.3. Let $C$ be a cumulative and structural inference operation on $\mathcal{L}$. Show that if for some finite $X_f \subseteq L, C(X_f) = L$, then $C$ is proper.

8.4. Let $M = \langle \mathcal{A}, \mathcal{D}, \mathcal{H}, \sqsubset \rangle$ be a stoppered preferential matrix such that $\langle \mathcal{A}, \mathcal{D} \rangle$ is proper. Let $N = \langle \mathcal{A}, \mathcal{D}, \mathcal{H} \rangle$. Show that for every $X \subseteq L$,

$$C_M(X) = L \textit{ iff } C_N(X) = L.$$

8.5. Let us call a stoppered preferential matrix $\langle \mathcal{A}, \mathcal{D}, \mathcal{H}, \sqsubset \rangle$ *special*, if the following conditions are satisfied:

- the matrix $\langle \mathcal{A}, \mathcal{D} \rangle$ is finite and proper;
- the set $\{\langle h(p_0), \ldots, h(p_{n-1}) \rangle : h \in \mathcal{H}\}$ can be effectively constructed, where $n$ is the cardinality of $\langle \mathcal{A}, \mathcal{D} \rangle$;
- there is $h \in \mathcal{H}$ that maps $\mathcal{L}$ onto $\mathcal{A}$.

Show that there is an effective procedure to determine whether or not the inference operation defined by a special preferential matrix satisfies the consistency preservation property.

8.6. Show that every supraclassical inference operation which satisfies the refutational principle and the consistency preservation property is decidable.

8.7. For every inference operation $C$ on $\mathcal{L}$, let

$$Th(C) = \{C(X) : X \subseteq L\}$$

denote the set of all *theories* of $C$. It is well-known that if $C$ and $C^*$ are two consequence operations, then $Th(C) = Th(C^*)$ implies $C = C^*$ (cf. [79], Section 1.2.3). However, this fact is not generally true for arbitrary inference operations. Show that if $C$ and $C^*$ are cumulative inference operations on $\mathcal{L}$, then

$$Th(C) = Th(C^*) \text{ implies } C_B = C^*_B.$$

8.8. The definitions of a decidable logic and of the refutational principle introduced in Section 1.1 can be adopted to inference systems without any change. Show that Theorem 1.2 remains true when the term 'logic' is replaced by 'inference system'.

# A

# RESOLUTION COUNTERPART $Rs_2$ OF $\mathcal{P}_2$

Classical propositional logic $\mathcal{P}_2$ is defined in Example 1.6. The resolution counterpart $Rs_2 = \langle \mathcal{V}_2, \mathcal{F}_2 \rangle$ of $\mathcal{P}_2$ is defined as follows.

**Verifiers of $Rs_2$:**

$$v_1 = p_0 \rightarrow p_0,$$
$$v_0 = \neg v_1.$$

**Transformation rules of $Rs_2$:**

$$\neg v_i \Rightarrow v_{1-i},$$
$$v_i \vee v_j \Rightarrow v_{max(i,j)},$$
$$v_i \wedge v_j \Rightarrow v_{min(i,j)},$$

$$v_i \rightarrow v_j \Rightarrow \begin{cases} v_0, & \text{if } i = 1 \text{ and } j = 0, \\ v_1, & \text{otherwise.} \end{cases}$$

**Family $\mathcal{F}_2$:** $\mathcal{F}_2 = \{\{v_0\}, \{v_0, v_1\}\}$.

**Resolution rule of $Rs_2$:**

$$\frac{\alpha_0(p), \alpha_1(p)}{\alpha_0(p/v_0) \vee \alpha_1(p/v_1)}.$$

**Termination condition**: for every finite $X \subseteq L$ clean in $Rs_2$,

$$C_2(X) = L \textit{ iff } v_0 \in r^*(X).$$

**Operator polarity and** $vo$**-polarity**:

$\neg$ is negative over its argument;

$\vee$ and $\wedge$ are positive over both their arguments;

$\rightarrow$ is negative over its first and positive over its second argument.

$R^+$ **and** $R^-$ **required by (pol2)**:

$$R^+\colon \frac{\alpha(p^{\oplus})}{\alpha(p/v_1)}, \quad R^-\colon \frac{\alpha(p^{\ominus})}{\alpha(p/v_0)}.$$

# B

# RESOLUTION COUNTERPART $Rs_3$ OF $\mathcal{P}_3$

The 3-valued Łukasiewicz logic $\mathcal{P}_3$ is defined in Example 1.6. The resolution counterpart $Rs_3 = \langle \mathcal{V}_3, \mathcal{F}_3 \rangle$ of $\mathcal{P}_3$ is defined as follows.

**Verifiers of $Rs_3$:**

$v_5 = p_0 \to p_0$, $v_4 = p_0 \vee \neg p_0$, $v_3 = \neg(v_4 \to \neg v_4)$,
$v_2 = \neg v_3$, $v_1 = \neg v_4$, $v_0 = \neg v_5$.

**Transformation rules of $Rs_3$:**

| $\to$ | 0 | 1 | 2 | 3 | 4 | 5 |
|---|---|---|---|---|---|---|
| 0 | 5 | 5 | 5 | 5 | 5 | 5 |
| 1 | 4 | 5 | 5 | 4 | 5 | 5 |
| 2 | 3 | 4 | 5 | 3 | 4 | 5 |
| 3 | 2 | 2 | 2 | 5 | 5 | 5 |
| 4 | 1 | 2 | 2 | 4 | 5 | 5 |
| 5 | 0 | 1 | 2 | 3 | 4 | 5 |

| $\vee$ | 0 | 1 | 2 | 3 | 4 | 5 |
|---|---|---|---|---|---|---|
| 0 | 0 | 1 | 2 | 3 | 4 | 5 |
| 1 | 1 | 1 | 2 | 4 | 4 | 5 |
| 2 | 2 | 2 | 2 | 5 | 5 | 5 |
| 3 | 3 | 4 | 5 | 3 | 4 | 5 |
| 4 | 4 | 4 | 5 | 4 | 4 | 5 |
| 5 | 5 | 5 | 5 | 5 | 5 | 5 |

| $\wedge$ | 0 | 1 | 2 | 3 | 4 | 5 |
|---|---|---|---|---|---|---|
| 0 | 0 | 0 | 0 | 0 | 0 | 0 |
| 1 | 0 | 1 | 1 | 0 | 1 | 1 |
| 2 | 0 | 1 | 2 | 0 | 1 | 2 |
| 3 | 0 | 0 | 0 | 3 | 3 | 3 |
| 4 | 0 | 1 | 1 | 3 | 4 | 4 |
| 5 | 0 | 1 | 2 | 3 | 4 | 5 |

| | $\neg$ |
|---|---|
| 0 | 5 |
| 1 | 4 |
| 2 | 3 |
| 3 | 2 |
| 4 | 1 |
| 5 | 0 |

**Family** $\mathcal{F}_3$: $\mathcal{F}_3 = \{V \subseteq Ver_{Rs_3} : V \not\subseteq \{v_2, v_5\}$ *and* $V \not\subseteq \{v_3, v_4, v_5\}\}$.

**Resolution rule of** $Rs_3$:

$$\frac{\alpha_0(p), \ldots, \alpha_5(p)}{\alpha_0(p/v_0) \vee \ldots \vee \alpha_5(p/v_5)}.$$

**Restricted resolution rule of** $Rs_3$:

$$\frac{\alpha_0(p), \alpha_1(p), \alpha_2(p)}{\alpha_0(p/v_3) \vee \alpha_1(p/v_4) \vee \alpha_2(p/v_5)}.$$

**Termination condition:** for every finite $X \subseteq L$ clean in $Rs_3$,

$$C_3(X) = L \textit{ iff } r^*(X) \cap \{v_0, v_1, v_3, v_4\} \neq \emptyset.$$

**Operator polarity**:

$\neg$ and $\rightarrow$ have no polarity values over their arguments;

$\vee$ and $\wedge$ are positive over both their arguments.

*vo*-**polarity** with respect to $\prec$ defined by the graph in Figure 5.3:

$\neg$ is negative over its argument;

$\vee$ and $\wedge$ are positive over both their arguments;

$\rightarrow$ is negative over its first and positive over its second argument.

$R^+$ **and** $R^-$ **required by (pol2):**

$$R^+: \frac{\alpha(p^{\oplus})}{\alpha(p/v_5)}, \quad R^-: \frac{\alpha(p^{\ominus})}{\alpha(p/v_0)}.$$

# C

# RESOLUTION COUNTERPART $Rs_4$ OF $\mathcal{P}_4$

The 4-valued Łukasiewicz logic $\mathcal{P}_4$ is defined in Example 1.6. The resolution counterpart $Rs_4 = \langle \mathcal{V}_4, \mathcal{F}_4 \rangle$ of $\mathcal{P}_4$ is defined as follows.

**Verifiers of $Rs_4$:**

$$v_7 = p_0 \to p_0 \ , v_6 = p_0 \vee \neg p_0 \ , v_5 = \neg(p_0 \to \neg p_0) \wedge (\neg p_0 \to p_0),$$
$$v_4 = \neg(p_0 \to (p_0 \to \neg p_0)) \wedge (\neg p_0 \to (\neg p_0 \to p_0)),$$
$$v_3 = \neg v_4 \ , v_2 = \neg v_5 \ , v_1 = \neg v_6 \ , v_0 = \neg v_7 .$$

**Transformation rules of $Rs_4$:**

| $\vee$ | 0 | 1 | 2 | 3 | 4 | 5 | 6 | 7 |
|---|---|---|---|---|---|---|---|---|
| 0 | 0 | 1 | 2 | 3 | 4 | 5 | 6 | 7 |
| 1 | 1 | 1 | 2 | 3 | 5 | 5 | 6 | 7 |
| 2 | 2 | 2 | 2 | 3 | 6 | 6 | 6 | 7 |
| 3 | 3 | 3 | 3 | 3 | 7 | 7 | 7 | 7 |
| 4 | 4 | 5 | 6 | 7 | 4 | 5 | 6 | 7 |
| 5 | 5 | 5 | 6 | 7 | 5 | 5 | 6 | 7 |
| 6 | 6 | 6 | 6 | 7 | 6 | 6 | 6 | 7 |
| 7 | 7 | 7 | 7 | 7 | 7 | 7 | 7 | 7 |

| $\wedge$ | 0 | 1 | 2 | 3 | 4 | 5 | 6 | 7 |
|---|---|---|---|---|---|---|---|---|
| 0 | 0 | 0 | 0 | 0 | 0 | 0 | 0 | 0 |
| 1 | 0 | 1 | 1 | 1 | 0 | 1 | 1 | 1 |
| 2 | 0 | 1 | 2 | 2 | 0 | 1 | 2 | 2 |
| 3 | 0 | 1 | 2 | 3 | 0 | 1 | 2 | 3 |
| 4 | 0 | 0 | 0 | 0 | 4 | 4 | 4 | 4 |
| 5 | 0 | 1 | 1 | 1 | 4 | 5 | 5 | 5 |
| 6 | 0 | 1 | 2 | 2 | 4 | 5 | 6 | 6 |
| 7 | 0 | 1 | 2 | 3 | 4 | 5 | 6 | 7 |

| $\to$ | 0 | 1 | 2 | 3 | 4 | 5 | 6 | 7 |
|---|---|---|---|---|---|---|---|---|
| 0 | 7 | 7 | 7 | 7 | 7 | 7 | 7 | 7 |
| 1 | 6 | 7 | 7 | 7 | 6 | 7 | 7 | 7 |
| 2 | 5 | 6 | 7 | 7 | 5 | 6 | 7 | 7 |
| 3 | 4 | 5 | 6 | 7 | 4 | 5 | 6 | 7 |
| 4 | 3 | 3 | 3 | 3 | 7 | 7 | 7 | 7 |
| 5 | 2 | 3 | 3 | 3 | 6 | 7 | 7 | 7 |
| 6 | 1 | 2 | 3 | 3 | 5 | 6 | 7 | 7 |
| 7 | 0 | 1 | 2 | 3 | 4 | 5 | 6 | 7 |

| | $\neg$ |
|---|---|
| 0 | 7 |
| 1 | 6 |
| 2 | 5 |
| 3 | 4 |
| 4 | 3 |
| 5 | 2 |
| 6 | 1 |
| 7 | 0 |

**Family $\mathcal{F}_4$**: $\mathcal{F}_4 = \{V \subseteq Ver_{Rs_4} : V \not\subseteq \{v_3, v_7\} \text{ and } V \not\subseteq \{v_4, v_5, v_6, v_7\}\}$.

**Resolution rule of $Rs_4$**:

$$\frac{\alpha_0(p), \ldots, \alpha_7(p)}{\alpha_0(p/v_0) \vee \ldots \vee \alpha_7(p/v_7)}.$$

**Restricted resolution rule of $Rs_4$**:

$$\frac{\alpha_0(p), \alpha_1(p), \alpha_2(p), \alpha_3(p)}{\alpha_0(p/v_4) \vee \alpha_1(p/v_5) \vee \alpha_2(p/v_6) \vee \alpha_3(p/v_7)}.$$

**Termination condition**: for every finite $X \subseteq L$ clean in $Rs_4$,

$$C_4(X) = L \quad \textit{iff} \quad r^*(X) \cap \{v_0, v_1, v_2, v_4, v_5, v_6\} \neq \emptyset.$$

**Operator polarity**:

$\neg$ and $\to$ have no polarity values over their arguments;

$\vee$ and $\wedge$ are positive over both their arguments.

*vo*-**polarity** with respect to $\prec$ defined by:

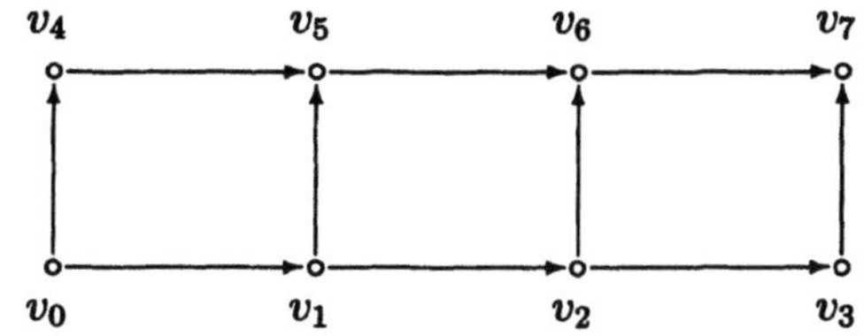


**$\neg$ is negative over its argument;**

**$\vee$ and $\wedge$ are positive over both their arguments;**

**$\rightarrow$ is negative over its first and positive over its second argument.**

**$R^+$ and $R^-$ required by (pol2):**

$$R^+ : \frac{\alpha(p^{\oplus})}{\alpha(p/v_7)}, \quad R^- : \frac{\alpha(p^{\ominus})}{\alpha(p/v_0)}.$$

# D

# RESOLUTION COUNTERPART $Rs_5$ OF $\mathcal{P}_5$

The 5-valued Łukasiewicz logic $\mathcal{P}_5$ is defined in Example 1.6. The resolution counterpart $Rs_5 = \langle \mathcal{V}_5, \mathcal{F}_5 \rangle$ of $\mathcal{P}_5$ is defined as follows.

**Verifiers of $Rs_5$:**

$v_0 = \neg(p_0 \to p_0)$, $v_1 = \neg(H_4 \to H_3)$, $v_8 = \neg v_1$,
$v_2 = v_8 \to v_1$, $v_3 = v_8 \to v_2$, $v_4 = v_8 \to v_3$,
$v_5 = \neg v_4$, $v_6 = \neg v_3$, $v_7 = \neg v_2$,
$v_9 = \neg v_0$,

where $H_3 = p_0 \to (p_0 \to \neg p_0)$ and $H_4 = p_0 \to H_3$.

**Transformation rules of $Rs_5$:**

| $\to$ | 0 | 1 | 2 | 3 | 4 | 5 | 6 | 7 | 8 | 9 |
|---|---|---|---|---|---|---|---|---|---|---|
| 0 | 9 | 9 | 9 | 9 | 9 | 9 | 9 | 9 | 9 | 9 |
| 1 | 8 | 9 | 9 | 9 | 9 | 8 | 9 | 9 | 9 | 9 |
| 2 | 7 | 8 | 9 | 9 | 9 | 7 | 8 | 9 | 9 | 9 |
| 3 | 6 | 7 | 8 | 9 | 9 | 6 | 7 | 8 | 9 | 9 |
| 4 | 5 | 6 | 7 | 8 | 9 | 5 | 6 | 7 | 8 | 9 |
| 5 | 4 | 4 | 4 | 4 | 4 | 9 | 9 | 9 | 9 | 9 |
| 6 | 3 | 4 | 4 | 4 | 4 | 8 | 9 | 9 | 9 | 9 |
| 7 | 2 | 3 | 4 | 4 | 4 | 7 | 8 | 9 | 9 | 9 |
| 8 | 1 | 2 | 3 | 4 | 4 | 6 | 7 | 8 | 9 | 9 |
| 9 | 0 | 1 | 2 | 3 | 4 | 5 | 6 | 7 | 8 | 9 |

| $\vee$ | 0 | 1 | 2 | 3 | 4 | 5 | 6 | 7 | 8 | 9 |
|---|---|---|---|---|---|---|---|---|---|---|
| 0 | 0 | 1 | 2 | 3 | 4 | 5 | 6 | 7 | 8 | 9 |
| 1 | 1 | 1 | 2 | 3 | 4 | 6 | 6 | 7 | 8 | 9 |
| 2 | 2 | 2 | 2 | 3 | 4 | 7 | 7 | 7 | 8 | 9 |
| 3 | 3 | 3 | 3 | 3 | 4 | 8 | 8 | 8 | 8 | 9 |
| 4 | 4 | 4 | 4 | 4 | 4 | 9 | 9 | 9 | 9 | 9 |
| 5 | 5 | 6 | 7 | 8 | 9 | 5 | 6 | 7 | 8 | 9 |
| 6 | 6 | 6 | 7 | 8 | 9 | 6 | 6 | 7 | 8 | 9 |
| 7 | 7 | 7 | 7 | 8 | 9 | 7 | 7 | 7 | 8 | 9 |
| 8 | 8 | 8 | 8 | 8 | 9 | 8 | 8 | 8 | 8 | 9 |
| 9 | 9 | 9 | 9 | 9 | 9 | 9 | 9 | 9 | 9 | 9 |

| $v$ | 0 | 1 | 2 | 3 | 4 | 5 | 6 | 7 | 8 | 9 |
|---|---|---|---|---|---|---|---|---|---|---|
| $\neg v$ | 9 | 8 | 7 | 6 | 5 | 4 | 3 | 2 | 1 | 0 |

| $\wedge$ | 0 | 1 | 2 | 3 | 4 | 5 | 6 | 7 | 8 | 9 |
|---|---|---|---|---|---|---|---|---|---|---|
| 0 | 0 | 0 | 0 | 0 | 0 | 0 | 0 | 0 | 0 | 0 |
| 1 | 0 | 1 | 1 | 1 | 1 | 0 | 1 | 1 | 1 | 1 |
| 2 | 0 | 1 | 2 | 2 | 2 | 0 | 1 | 2 | 2 | 2 |
| 3 | 0 | 1 | 2 | 3 | 3 | 0 | 1 | 2 | 3 | 3 |
| 4 | 0 | 1 | 2 | 3 | 4 | 0 | 1 | 2 | 3 | 4 |
| 5 | 0 | 0 | 0 | 0 | 0 | 5 | 5 | 5 | 5 | 5 |
| 6 | 0 | 1 | 1 | 1 | 1 | 5 | 6 | 6 | 6 | 6 |
| 7 | 0 | 1 | 2 | 2 | 2 | 5 | 6 | 7 | 7 | 7 |
| 8 | 0 | 1 | 2 | 3 | 3 | 5 | 6 | 7 | 8 | 8 |
| 9 | 0 | 1 | 2 | 3 | 4 | 5 | 6 | 7 | 8 | 9 |

**Family $\mathcal{F}_5$:** $\mathcal{F}_5 = \{V \subseteq Ver_{Rs_5} : V \not\subseteq \{v_4, v_9\}$ *and* $V \not\subseteq \{v_5, \ldots, v_9\}\}$.

**Resolution rule of $Rs_5$:**

$$\frac{\alpha_0(p), \ldots, \alpha_9(p)}{\alpha_0(p/v_0) \vee \ldots \vee \alpha_9(p/v_9)}.$$

**Restricted resolution rule of $Rs_5$:**

$$\frac{\alpha_0(p), \alpha_1(p), \alpha_2(p), \alpha_3(p), \alpha_4(p)}{\alpha_0(p/v_5) \vee \alpha_1(p/v_6) \vee \alpha_2(p/v_7) \vee \alpha_3(p/v_8) \vee \alpha_4(p/v_9)}.$$

**Termination condition**: for every finite $X \subseteq L$ clean in $Rs_5$,

$$C_5(X) = L \quad \textit{iff} \quad r^*(X) \cap \{v_0, \ldots, v_3, v_5, \ldots, v_8\} \neq \emptyset.$$

**Operator polarity:**

$\neg$ and $\rightarrow$ have no polarity values over their arguments;

$\vee$ and $\wedge$ are positive over both their arguments.

$vo$-**polarity** with respect to $\prec$ defined by:

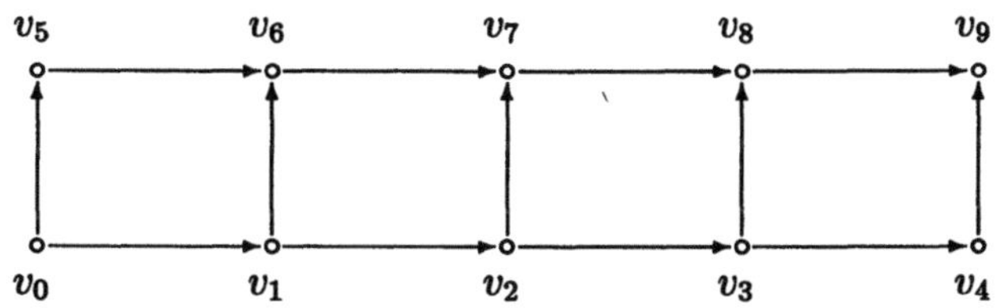


$\neg$ is negative over its argument;

$\vee$ and $\wedge$ are positive over both their arguments;

$\rightarrow$ is negative over its first and positive over its second argument.

$R^+$ **and** $R^-$ **required by (pol2):**

$$R^+\colon \frac{\alpha(p^{\oplus})}{\alpha(p/v_9)}, \quad R^- : \frac{\alpha(p^{\ominus})}{\alpha(p/v_0)}.$$

# E

# RESOLUTION COUNTERPART $Rs_3^*$ OF $\mathcal{P}_3^*$

The 3-valued Post logic $\mathcal{P}_3^* = \langle \mathcal{L}, C_3^* \rangle$ is defined in Example 1.6. The resolution counterpart $Rs_3^* = \langle \mathcal{V}_3^*, \mathcal{F}_3^* \rangle$ for $\mathcal{P}_3^*$ is defined as follows.

**Verifiers of $Rs_3^*$:**

$$v_2 = p_0 \to p_0, \quad v_1 = \neg v_2, \quad v_0 = \neg v_1.$$

**Transformation rules of $Rs_3^*$:**

$$\neg v_0 \Rightarrow v_2, \neg v_1 \Rightarrow v_0, \neg v_2 \Rightarrow v_1,$$
$$v_i \vee v_j \Rightarrow v_{max(i,j)},$$
$$v_i \wedge v_j \Rightarrow v_{min(i,j)},$$

$$v_i \to v_j \Rightarrow \begin{cases} v_2, & \text{if } i \leq j, \\ v_j, & \text{otherwise.} \end{cases}$$

**Family $\mathcal{F}_3^*$:** $\mathcal{F}_3^* = \{V \subseteq \{v_0, v_1, v_2\} : V \neq \{v_2\}\}$.

**Resolution rule of $Rs_3^*$:**

$$\frac{\alpha_0(p), \alpha_1(p), \alpha_2(p)}{\alpha_0(p/v_0) \vee \alpha_1(p/v_1) \vee \alpha_2(p/v_2)}.$$

**Termination condition:** for every finite $X \subseteq L$ clean in $Rs_3^*$,

$$C_3^*(X) = L \textit{ iff } v_0 \in r^*(X) \text{ or } v_1 \in r^*(X).$$

**Operator polarity**:

$\neg$ and $\rightarrow$ have no polarity values over their arguments;

$\vee$ and $\wedge$ are positive over both their arguments.

*vo*-**polarity** with respect to $\prec$ defined by:

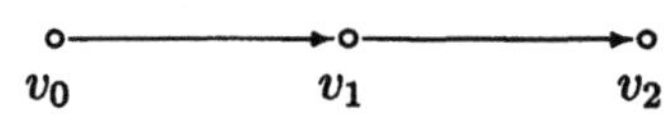


$\neg$ is negative over its argument;

$\vee$ and $\wedge$ are positive over both their arguments;

$\rightarrow$ is negative over its first and positive over its second argument.

$R^+$ **and** $R^-$ **required by (pol2)**:

$$R^+\colon \frac{\alpha(p^{\oplus})}{\alpha(p/v_2)}, \quad R^-\colon \frac{\alpha(p^{\ominus})}{\alpha(p/v_0)}.$$

# F

# RESOLUTION COUNTERPART $Rs_3^S$ OF $\mathcal{P}_S$

The 3-valued Słupecki logic $\mathcal{P}_s = \langle \mathcal{L}_s, C_s \rangle$ is defined in Example 4.9. The resolution counterpart $Rs_3^s = \langle \mathcal{V}_s, \mathcal{F}_s \rangle$ of $\mathcal{P}_s$ is defined as follows.

**Verifiers of $Rs_3^s$:**

$$v_2 = \mathbf{1} \rightarrow \mathbf{1}, \quad v_1 = \mathbf{1}, \quad v_0 = \neg v_2.$$

**Constants of $\mathcal{V}_s$: 1.**

**Transformation rules of $Rs_3^s$:**

| $\vee$ | 0 | 1 | 2 |
|---|---|---|---|
| 0 | 0 | 1 | 2 |
| 1 | 1 | 1 | 2 |
| 2 | 2 | 2 | 2 |

| $\wedge$ | 0 | 1 | 2 |
|---|---|---|---|
| 0 | 0 | 0 | 0 |
| 1 | 0 | 1 | 1 |
| 2 | 0 | 1 | 2 |

| $\rightarrow$ | 0 | 1 | 2 |
|---|---|---|---|
| 0 | 2 | 2 | 2 |
| 1 | 1 | 2 | 2 |
| 2 | 0 | 1 | 2 |

| | $\neg$ |
|---|---|
| 0 | 2 |
| 1 | 1 |
| 2 | 0 |

**Family $\mathcal{F}_s$:** $\mathcal{F}_s = \{V \subseteq \{v_0, v_1, v_2\} : V \neq \{v_2\}\}.$

**Resolution rule of $Rs_3^s$:**

$$\frac{\alpha_0(p), \alpha_1(p), \alpha_2(p)}{\alpha_0(p/v_0) \vee \alpha_1(p/v_1) \vee \alpha_2(p/v_2)}.$$

**Termination condition**: for every finite $X \subseteq L_s$ clean in $Rs^s_3$,

$$C_s(X) = L_s \;\; \textit{iff} \;\; v_0 \in r^*(X) \text{ or } v_1 \in r^*(X).$$

**Operator polarity**:

$\neg$ and $\rightarrow$ have no polarity values over their arguments;

$\vee$ and $\wedge$ are positive over both their arguments.

*vo*-**polarity** with respect to $\prec$ defined by:

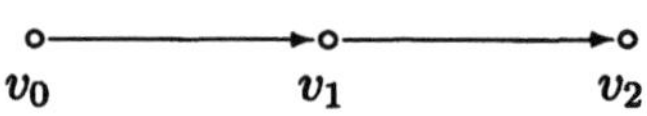


$\neg$ is negative over its argument;

$\vee$ and $\wedge$ are positive over both their arguments;

$\rightarrow$ is negative over its first and positive over its second argument.

$R^+$ **and** $R^-$ **required by (pol2)**:

$$R^+: \frac{\alpha(p^{\oplus})}{\alpha(p/v_2)}, \quad R^-: \frac{\alpha(p^{\ominus})}{\alpha(p/v_0)}.$$

# REFERENCES

The abbreviations of names of serials which are used in References are adopted from *Mathematical Reviews*.


[1] M. Abadi and Z. Manna, Modal Theorem Proving, in *Proc. 8th Int. Conf. on Automated Deduction*, Oxford (1986), *Lecture Notes in Comput. Sci.* **230**, 172–189.

[2] M. Bazz and C. G. Fermüller, Resolution for Many-Valued Logics, in *Proc. Int. Conf. on Logic Programming and Automated Reasoning*, St. Petersburg (1992), *Lecture Notes in Artif. Intell.* **624**, 107–118.

[3] M. Bazz, C. G. Fermüller and R. Zach, Systematic Construction of Natural Deduction Systems for Many-Valued Logics, in *Proc. Twenty-Third Int. Symp. on Multiple-Valued Logic*, Sacramento (1993), 208–213.

[4] G. Beavers, Automated Theorem Proving for Łukasiewicz Logics, *Studia Logica* **52** (1993), 183–196.

[5] L. Bolc and P. Borowik, *Many-Valued Logics. 1: Theoretical Foundations,* Springer-Verlag (1992).

[6] A. L. Brown and Y. Shoham, New Results on Semantical Nonmonotonic Reasoning, in *Proc. 2nd Int. Workshop on Non-Monotonic Reasoning*, Grassau (1988), *Lecture Notes in Artif. Intell.* **346**, 19–26.

[7] C. Chang and R. C. Lee, *Symbolic Logic and Mechanical Theorem Proving*, Academic Press (1973).

[8] J. Chazarain, A. Riscos, J. A. Alonso and E. Briales, Multi-valued Logic and Gröbner Bases with Applications to Modal Logic, *J. Symbolic Comput.* **11** (1991), 181–194.

[9] L. Couturat, *L'algèbre de la Logique*, Paris (1905).

[10] J. Czelakowski, *Model-Theoretic Methods in Methodology of Propositional Calculi*, Polish Academy of Sciences (1980).

[11] P. Doherty and W. Łukaszewicz, NML3 – A Non-monotonic Logic with Explicit Defaults, *J. of Appl. Non-Classical Logics* **2** (1992), 9–48.

[12] M. Fitting, *Proof Methods for Modal and Intuitionistic Logics*, D. Reidel, Dordrecht (1983).

[13] M. Fitting, First-Order Modal Tableaux, *J. Automat. Reasoning* **4** (1988), 191–213.

[14] A. M. Frisch and R. B. Scherl, A General Framework for Modal Deduction, in *Proc. Second Int. Conf. on Principles of Knowledge Representation and Reasoning*, San Mateo (1991), Morgan Kaufmann.

[15] D. M. Gabbay, Theoretical Foundations for Non-Monotonic Reasoning in Expert Systems, in *Logics and Models of Concurrent Systems* (K. Apt ed.), 439–457, Springer-Verlag (1985).

[16] I. P. Gent, Theory Matrices (for Modal Logics) Using Alphabetical Monotonicity, *Studia Logica* **52** (1993), 233–257.

[17] G. Grätzer, *Universal Algebra* (2nd edition), Springer-Verlag (1979).

[18] E. Harley and Z. Stachniak, Minimal Resolution Proof Systems for Finitely-Valued Łukasiewicz Logics, in *Proc. Twenty-Third Int. Symp. on Multiple-Valued Logic*, Sacramento (1993), 54–59.

[19] R. Hähnle, *Automated Deduction in Multiple-valued Logics*, Oxford University Press (1993).

[20] J. Hawranek and M. Tokarz, Matrices for Predicate Logics, *Bull. Sect. Logic Polish Academy of Sciences* **6** (1977), 27–31.

[21] J. E. Hopcroft and J. D. Ullman, *Introduction to Automata Theory, Languages, and Computation*, Addison-Wesley (1979).

[22] K. Konolige, Resolution and Quantified Epistemic Logics, in *Proc. 8th Int. Conf. on Automated Deduction*, Oxford (1986), *Lecture Notes in Comput. Sci.* **230**, 199–208.

[23] S. Kraus, D. Lehmann and M. Magidor, Nonmonotonic Reasoning, Preferential Models and Cumulative Logics, *Artificial Intelligence* **44** (1990), 167–207.

[24] J. J. Lu and L. J. Henschen, The Completeness of gp-Resolution for Annotated Logics, *Inform. Process. Lett.* **44** (1992), 135–140.

[25] J. Łoś and R. Suszko, Remarks on Sentential Logics, *Indagationes Mathematicae* **20** (1958), 177–183.

[26] J. Łukasiewicz, Logika dwuwartościowa (Two-Valued Logic), *Przegląd Filozoficzny* **23** (1921), 189–205. (English translation in J. Łukasiewicz, *Selected Works*, L. Borkowski, ed., North-Holland (1970).)

[27] D. Makinson, General Theory of Cumulative Inference, in *Proc. 2nd Int. Workshop on Non-Monotonic Reasoning*, Grassau (1988), *Lecture Notes in Artif. Intell.* **346**, 1–18.

[28] D. Makinson, General Patterns in Nonmonotonic Reasoning, in *Handbook of Logic in Artificial Intelligence and Logic Programming*, Vol 3, *Nonmonotonic Reasoning and Uncertain Reasoning* (D. M. Gabbay, C. J. Hogger, J. A. Robinson and D. Nute, eds.), 35–110, Oxford University Press (1994).

[29] G. Malinowski, *Many-Valued Logics*, Oxford University Press (1993).

[30] Z. Manna and R. Waldinger, Special Relations in Automated Deduction, *J. Assoc. Comput. Mach.* **33** (1986), 1–59.

[31] V. W. Marek and M. Truszczyński, *Nonmonotonic Logic: Context-Dependent Reasoning*, Springer-Verlag (1993).

[32] A. Martelli and U. Montanari, An Efficient Unification Algorithm, *ACM Trans. Programming Languages and Systems* **4** (1982), 258–282.

[33] J. McCharen, R. Overbeek and L. Wos, Problems and Experiments for and with Automated Theorem Proving Programs, *IEEE Trans. Comput.* **C-25** (1976), 773–782.

[34] D. McDermott, Nonmonotonic Logic II: Nonmonotonic Modal Theories, *J. Assoc. Comput. Mach.* **29** (1982), 33–57.

[35] R. McNaughton, A Theorem About Infinite-Valued Sentential Logic. *J. Symbolic Logic* **16** (1951), 1–13.

[36] R. C. Moore, Semantical Considerations on Nonmonotonic Logic, *Artificial Intelligence* **25** (1985), 75–94.

[37] C. G. Morgan, Methods for Automated Theorem Proving in Nonclassical Logics, *IEEE Trans. Comput.* **C-25** (1976), 852–862.

[38] C. G. Morgan, A resolution Principle for a Class of Many-Valued Logics, *Logique et Analyse* **74-75-76** (1976), 311–339.

[39] N. V. Murray, Completely Non-Clausal Theorem Proving, *Artificial Intelligence* **18** (1982), 67–85.

[40] N. V. Murray and E. Rosenthal, Resolution and Path Dissolution in Multiple-Valued Logics, in *Proc. 6th Int. Symp. on Methodologies for Intelligent Systems*, Charlotte (1991), *Lecture Notes in Artif. Intell.* **542**, 570–579.

[41] N. V. Murray and E. Rosenthal, Dissolution: Making Paths Vanish, *J. Assoc. Comput. Mach.* **40** (1993), 504–535.

[42] B. H. Neumann, *Universal Algebra, Lecture Notes*, New York (1962).

[43] P. W. O'Hearn and Z. Stachniak, Resolution Framework for Finitely-Valued First-Order Logics, *J. Symbolic Comput.* **13** (1992), 235–254.

[44] H. J. Ohlbach, A Resolution Calculus for Modal Logics, in *Proc. 9th Int. Conf. on Automated Deduction*, Argonne (1988), *Lecture Notes in Comput. Sci.* **310**, 500–516.

[45] E. Orłowska, Resolution Systems and Their Applications II, *Fundamenta Informaticae* **3** (1980), 333–362.

[46] E. Orłowska, Mechanical Proof Methods for Post Logics, *Logique et Analyse* **110–111** (1985), 173–192.

[47] D. Pearce, Default Logic and Constructive Logic, in *Proc. 10th European Conf. on Artificial Intelligence*, Vienna (1992), 309–313, John Wiley.

[48] F. J. Pelletier, Identity in Modal Logic Theorem Proving, *Studia Logica* **52** (1993), 291–308.

[49] B. Piochi, Logical Matrices and Non-Structural Consequence Operators, *Studia Logica* **42** (1983), 33–42.

[50] D. Poole, A Logical Framework for Default Reasoning, *Artificial Intelligence* **36** (1988), 27–47.

[51] E. L. Post, Introduction to a General Theory of Elementary Propositions, *Amer. J. Math.* **43** (1921), 163–185.

[52] H. Rasiowa, *An Algebraic Approach to Non-Classical Logics*, North-Holland–PWN (1974).

[53] N. Rescher, *Many-Valued Logics*, McGraw-Hill (1969).

[54] J. A. Robinson, A Machine-Oriented Logic Based on the Resolution Principle, *J. Assoc. Comput. Mach.* **12** (1965), 23–41.

[55] J. B. Rosser and A. R. Turquette, *Many-Valued Logics*, North-Holland (1952).

[56] P. H. Schmitt, Computational Aspects of Three-Valued Logic, in *Proc. 8th Int. Conf. on Automated Deduction*, Oxford (1986), *Lecture Notes in Comput. Sci.* **230**, 190–198.

[57] E. Schröder, *Vorlesungen über die Algebra der Logik*, I Band, Leipzig (1890).

[58] W. Snyder, *A Proof Theory for General Unification*, Birkhäuser (1991).

[59] J. T. J. Srzednicki and Z. Stachniak, *Leśniewski's Lecture Notes in Logic*, Kluwer Academic Publishers (1988).

[60] Z. Stachniak, Two Theorems on Many-Valued Logics, *J. Philos. Logic* **17** (1988), 171–179.

[61] Z. Stachniak, Resolution Rule: An Algebraic Perspective, in *Proc. Algebraic Logic and Universal Algebra in Computer Science Conf.*, Ames (1988), *Lecture Notes in Comput. Sci.* **425**, 227–242.

[62] Z. Stachniak, Note on Constructibility of Resolution Proof Systems, in *Proc. European Workshop on Logics in AI*, Amsterdam (1990), *Lecture Notes in Artif. Intell.* **478**, 487–498.

[63] Z. Stachniak, Minimization of Resolution Proof Systems, *Fundamenta Informaticae* **14** (1991), 129–146.

[64] Z. Stachniak, Extending Resolution to Resolution Logics, *J. Expt. Theor. Artif. Intell.* **3** (1991), 17–32.

[65] Z. Stachniak, Note on Resolution Circuits, in *Proc. Int. Symp. on Methodologies for Intelligent Systems*, Charlotte (1991), *Lecture Notes in Artif. Intell.* **542**, 620–629.

[66] Z. Stachniak, Resolution Approximation of First-Order Logics, *Information and Computation* **96** (1992), 225-244.

[67] Z. Stachniak, An Essay on Resolution Logics, *Studia Logica* **52** (1993), 309–322.

[68] Z. Stachniak, Algebraic Semantics for Cumulative Inference Operations, in *Proc. AAAI-93*, Washington, DC (1993), 444–449, AAAI Press/MIT Press.

[69] Z. Stachniak, Lattices of Resolution Logics, in *Proc. 24th Int. Symp. on Multiple-Valued Logic*, Boston (1994), 108–114.

[70] Z. Stachniak, Nonmonotonic Theories and their Axiomatic Varieties. To appear in *J. Logic, Language and Information* (199?).

[71] Z. Stachniak and P. O'Hearn, Resolution in the Domain of Strongly Finite Logics, *Fundamenta Informaticae* **13** (1990), 333–351.

[72] A. Tarski, O wyrazie pierwotnym logistyki (On the Primitive Term of Logistic), *Przegląd Filozoficzny* **26** (1923), 68–89. (English translation in A. Tarski, *Logic, Semantics, Mathematics, Papers from 1923 to 1938*, translated by J. H. Woodger, Oxford University Press (1956).)

[73] P. B. Thistlewaite, M. A. McRobbie and R. K. Meyer, *Automated Theorem-Proving in Non-Classical Logics*, Pitman (1988).

[74] H. Tuominen, Dynamic Logic as a Uniform Framework for Theorem Proving in Intensional Logic, in *Proc. 10th Int. Conf. on Automated Deduction*, Kaiserslautern (1990), *Lecture Notes in Artif. Intell.* **449**, 514–527.

[75] L. A. Wallen, *Automated Proof Search in Non-Classical Logics*, The MIT Press (1990).

[76] J. Waszkiewicz and B. Węglorz, On Products of Structures for Generalized Logics, *Studia Logica* **25** (1969), 7–15.

[77] L. Wos, *Automated Reasoning: 33 Basic Research Problems*, Prentice Hall (1988).

[78] L. Wos, G. A. Robinson and D. F. Carson, Efficiency and Completeness of the Set of Support Strategy in Theorem Proving. *J. Assoc. Comput. Mach.* **12** (1965), 536–541.

[79] R. Wójcicki, *Theory of Logical Calculi: Basic Theory of Consequence Operations*, Kluwer Academic Publishers (1988).

[80] J. Zygmunt, Uwagi o Konsekwencjach Matrycowych (Polish), *Acta Universitatis Wratislaviensis, Logika* **8** (1983), 47–58.

[81] J. Zygmunt, An Application of the Lindenbaum Method in the Domain of Strongly Finite Sentential Calculi, *Acta Universitatis Wratislaviensis, Logika* **8** (1983), 59–68.

# INDEX

# INDEX OF SYMBOLS

## Automated Reasoning Series

1. R.S. Boyer (ed.): *Automated Reasoning*. Essays in Honor of Woody Bledsoe. 1991 ISBN 0-7923-1409-3
2. A. Quaife: *Automated Development of Fundamental Mathematical Theories*. 1992 ISBN 0-7923-2021-2
3. J Strother Moore: *Piton*. A Mechanically Verified Assembly-Level Language. 1996 ISBN 0-7923-3920-7
4. Z. Stachniak: *Resolution Proof Systems*. An Algebraic Theory. 1996 ISBN 0-7923-4017-5

KLUWER ACADEMIC PUBLISHERS – DORDRECHT / BOSTON / LONDON

MIX
Papier aus verantwortungsvollen Quellen
Paper from responsible sources
FSC® C105338

If you have any concerns about our products,
you can contact us on
**ProductSafety@springernature.com**

In case Publisher is established outside the EU,
the EU authorized representative is:
**Springer Nature Customer Service Center GmbH**
**Europaplatz 3, 69115 Heidelberg, Germany**

Printed by Libri Plureos GmbH
in Hamburg, Germany